Harvey Lomax

D1273084

HUMANS AND OTHER ANIMALS

Humans and Other Animals

JOHN DUPRÉ

CLARENDON PRESS · OXFORD

OXFORD
UNIVERSITY PRESS

Great Clarendon Street, Oxford OX2 6DP

Oxford University Press is a department of the University of Oxford.
It furthers the University's objective of excellence in research, scholarship,
and education by publishing worldwide in

Oxford New York

Auckland Bangkok Buenos Aires Cape Town Chennai
Dar es Salaam Delhi Florence Hong Kong Istanbul Karachi Kolkata
Kuala Lumpur Madrid Melbourne Mexico City Mumbai Nairobi
São Paulo Shanghai Singapore Taipei Tokyo Toronto
with an associated company in Berlin

Oxford is a registered trade mark of Oxford University Press
in the UK and in certain other countries

Published in the United States
By Oxford University Press Inc., New York

British Library Cataloguing in Publication Data

Data available

Library of Congress Cataloging in Publication Data

Data available

ISBN 0-19-924709-9

1 3 5 7 9 10 8 6 4 2

Typeset in 10.5/12 pt Minion
by Graphicraft Limited, Hong Kong
Printed in Great Britain by
Biddles Ltd., Guildford & King's Lynn

Preface

It is perhaps a little awkward for someone who has devoted a career to promoting the idea of knowledge as disconnected and local to claim that a collection of apparently disparate papers represents, nevertheless, a sufficiently unified body of work to merit bringing together as an anthology. Still, the task is not hopeless. My ideas about the disunity of science grew out of an appreciation of the complexities and local specificities of classification, and a concern with the problems of classification is the theme that unites these essays. Naturally these problems are somewhat diverse. The major subthemes are the classification of organisms, the classification of people, and the criteria by which we differentiate people from other organisms.

Six of the papers were written in the last few years, and therefore represent my latest thinking on these topics. Three essays, Chapters 1, 6, and 8, are considerably older. I include them because they are the occasions for my first fully articulated views on the central topics of biological kinds, human kinds, and essentialism. Though I would no doubt write them differently today, I find myself still basically in sympathy with what they say. Those who find my more recent positions profoundly misguided may look to these essays to see where I took the first fatal missteps.

The last two chapters are of intermediate age and are included for a somewhat different reason. The topics they address, animal minds and animal language, have mainly concerned rather different audiences from those to which the earlier papers are addressed. They have thus been somewhat disconnected from the body of thought out of which they grew. I hope they may attract the interest of some new readers here, as well as illustrating some rather different directions in which the main lines of my ideas on biology have seemed to me to lead.

All the reprinted papers are identical to their original versions with regard to content, with only minor stylistic changes or corrections to typographical errors.

Over the twenty years during which these essays have been written I have been helped by many more people than I can hope to thank or even remember. A few are mentioned in specific acknowledgements to the various essays. I hope that the rest will know who they are and excuse my apparent ingratitude. Here I thank specifically only Regenia Gagnier. We had not met when I wrote the first essay in the book. All the others are better than they would have been without her intellectual stimulation and companionship.

Acknowledgements

I am most grateful to the original publishers for permission to reprint the following papers:

'Natural Kinds and Biological Taxa', *Philosophical Review*, 90 (1981), 66–91. Copyright 1981 Cornell University. Reprinted by permission of the publisher.

'Are Whales Fish?', in D. L. Medin and S. Atran (eds.), *Folkbiology* (MIT Press, 1999), 461–76.

'On the Impossibility of a Monistic Account of Species', in Robert A. Wilson (ed.), *Species: New Interdisciplinary Essays* (Cambridge University Press, 1999), 3–20.

'In Defence of Classification', *Studies in the History and Philosophy of the Biological and Biomedical Sciences*, 32 (2001), 203–19.

'Is "Natural Kind" a Natural Kind Term?', *Monist*, 85 (2002), 29–49.

'Human Kinds', in J. Dupré (ed.), *The Latest on the Best* (Bradford Books/MIT Press, 1987), 327–48.

'Sex, Gender, and Essence', *Midwest Studies in Philosophy*, 11 (1986), 441–57. Reprinted by permission of Blackwell publishers.

'What the Theory of Evolution Can't Tell Us', *Critical Quarterly*, 42 (2000), 18–34. Reprinted by permission of Blackwell publishers.

'The Mental Lives of Non-Human Animals', in M. Bekoff and D. Jamieson (eds.), *Interpretation and Explanation in the Study of Behavior: Comparative Perspectives* (Westview Press, 1990), 428–48.

'Conversations with Apes: Reflections on the Scientific Study of Language', in J. Hyman (ed.), *Investigating Psychology: Sciences of the Mind after Wittgenstein* (Routledge & Kegan Paul, 1991), 95–116.

Contents

Introduction

This volume begins where my interest in its central topic began, with our everyday classification of animals. That central topic, which informs to some degree every chapter, is classification, the sorting of things into kinds. The first two sections concern the sorting of animals into kinds, the next two the sorting of humans into kinds, and the final section considers some of the main ways in which we are inclined to distinguish ourselves, as a kind, from non-human, 'lower', animals.

My use of the word 'everyday' in describing the topic of the first chapter was perhaps too quick. As is standard contemporary philosophical practice, I was thinking of a contrast between the quotidian and the scientific. But a central point of Chapter 1 is to suggest that this dichotomy is much too simple. Classification of organisms is carried out in diverse and more or less technical ways by amateur enthusiasts as well as those with professional but non-scientific concerns, such as foresters, herbalists, and chefs. This apparently simple point led me to serious scepticism about any sharp boundary between the scientific and the mundane, and from there to doubts whether there could be any universal criterion of the scientific. This was the topic of my earlier book *The Disorder of Things* (1993). The present essays trace my explorations of the more evident topic of this early paper, classification.

The catalyst for writing Chapter 1 was not primarily biology, however, but the philosophy of language. The hottest thing in philosophy at the time I began my professional career was what came to be called the new theory of reference, then being developed by philosophers such as Keith Donellan and Saul Kripke. One aspect of this theory was an account of natural kind terms gestured at by Kripke and developed in detail by Hilary Putnam. This generated a huge amount of discussion, much of it focused on the famous, or perhaps notorious, 'Twin Earth' examples developed by Putnam. What struck me was something rather different, namely that the discussion just assumed the essentialism about, among other things, biological kinds. This struck me as problematic in the first place merely drawing on my experience as an amateur botanist, a fortunate avocation as plants are much more recalcitrant to the essentialist's classificatory assumptions than are most kinds of animals. But also at this time I had begun to learn some real biology, mostly from Ernst Mayr's superb *Animal Species and Evolution* (1963). Then it became clear to me that the essentialist perspective was demonstrably hopeless for any properly post-Darwinian biology.

Putnam's account of natural kinds assumed that there were sharply distinguished kinds of things out in the world and that, starting with a sense of the most obvious of these and the intention that our kind terms should refer to real natural kinds, with the help of science we had gradually refined our ability to discriminate these kinds. The theory had some very attractive consequences. For example, the earlier assumption that the extension of a scientific term was determined by the theories held to be true of its referents led to the embarrassing consequence that every change in scientific belief was likely to change the referent of scientific terms. When Millikan was supposedly measuring the charge on an electron, he was not measuring the charge on anything contemporary scientists would refer to by the term 'electron'. Putnam's idea immediately disposed of this difficulty. 'Electron' was introduced to refer to a certain natural kind of things and has always referred to this kind. It is just that we have become better at describing the members of the kind.

Real stories, unfortunately, including the one about electrons, are a lot more complicated than this. Ian Hacking (1983) has made impressive efforts to salvage the aforementioned benefits of Putnam's idea from the complexities of real scientific history. My worry, however, was rather whether the world was properly populated with the right sort of natural kinds to allow the story to get off the ground. A fundamental distinction here is whether there are such kinds available to the informal interests of everyday speech, or whether, even if not, there may be natural kinds discernible by the more refined methods of science. Putnam's argument is centrally addressed to the first point. We identify the kinds in advance of science and science subsequently tells us about the inner nature of the things we have distinguished: first there was water, then science told us it was really H_2O.

The overwhelming majority of examples for which this story might make sense are from biology. There are no more than a handful of terms from chemistry that exist as both pre-scientific and scientific kind terms: gold and water, endlessly discussed in the literature; a few other metals; table salt. Many pre-scientific terms that might have been thought to have a simple chemical nature have disappointed: earth, air, sand, mud, rock, and so on. Biology, on the other hand, offers a host of potential natural kinds. Sadly, and this finally is the main message of Chapter 1, Putnam's story does not work for biology. As I show, many biological kinds in ordinary language map quite poorly on to the kinds of biological science. And where they do fit, they very typically fit on to kinds that are largely arbitrary from a scientific perspective. Moths, lilies, hawks, rabbits, ducks, frogs, or cedars, for instance, have no plausible and coherent correlate in scientific classification.

In that chapter I also raise doubts about the common essentialist interpretation of scientific classification itself in biology. There are hints of the pluralistic perspective on scientific classification developed in later work, but the details of this are addressed in the two papers in Part II. Chapter 2, on the other hand, explores further the relationship between scientific and everyday classification. The willingness of people to accept a story such as Putnam's is reflected in the readiness with which people have come to accept supposedly scientific revisions of their classificatory schemes. Every schoolchild now knows that whales are not fish, and that this discovery was a scientific achievement. My own view, elaborated in Chapter 2, is that far from a scientific discovery, this was a bit of wholly unmotivated linguistic legislation, and one well suited to perpetuating a range of false beliefs about biological kinds.

The actual discovery, of course, was that whales are mammals. The physiology of a whale is, in many respects, more like a hedgehog than a tuna or a shark, and it is more closely related, by descent, to the hedgehog. But why should this be taken to imply that whales are not fish, rather than that some mammals are fish? The answer, I take it, is that people believe that both mammals and fish are natural kinds, and as such form part of a divergent hierarchy of kinds such that no kinds overlap: any two kinds are either disjoint or one includes the other. This is indeed a principle of most biological taxonomies. But by no stretch of the imagination is 'fish' a term of such a taxonomy. An enormously diverse group of organisms are included under the category of fish, and I can imagine no convincing reason for excluding the cetaceans (whales and porpoises) from this broad church. In fact a salmon, say, is less distantly related to a porpoise than it is to an ancient and 'primitive' fish such as a lamprey.

Although in these chapters I oppose a certain kind of realism about natural kinds, I do not do so in defence of nominalism. It is not the reality I want to oppose so much as the exclusivity or elitism of the kinds that are allegedly provided by science. It is the idea that there is only one legitimate way of classifying things in the world, the 'scientific' way, that presumably underlies the assumption that whales must be either fish or mammals, but cannot be both. Yet 'fish' really means no more than 'aquatic vertebrate', and whales, sharks, and salmon are all aquatic vertebrates. And of course this doesn't mean that there is no such thing as fishkind. As far as I can see, the kind aquatic vertebrate is no less real than the kind mammal, although somewhat easier to gain admittance to. Once we follow the logic of Darwinism in disposing of the idea that an organism has an essence that determines its necessary place in a unique nested hierarchy of kinds, there is no reason to exclude the possibility of a variety of different classificatory schemes, suited to a variety of purposes, some scientific and some not, that may criss-cross and overlap one another in various ways. This picture, which I have called 'promiscuous realism', is developed in the first part of the book, and provides a fundamental motivation for the ideas developed in subsequent chapters as well.

A central point of the first two chapters was to insist on the independence of ordinary language from the taxonomies of science. The case will of course be much stronger if it turns out that science itself can provide no unique or privileged classification. That this is in fact the case is the main thesis defended, for biology at least, in Chapter 3. As is traditional in all theoretical work in biology, I claim that such a view is an inevitable consequence of a proper appreciation of Darwin. Though it remains a matter of controversy, an increasing number of philosophers and theoretically minded biologists have concluded that there is no universal principle by which organisms of all kinds can be sorted into species. Criteria that work well for mammals or birds have no application to flowering plants or protozoans. The absence of a unique and universal taxonomic scheme makes it natural to consider the likelihood that different principles of classification may be most appropriate not only for different groups of organisms, but also for different kinds of enquiry. Investigations into phylogeny will make very different demands from investigations into ecology, say, and may perfectly well call for different taxonomic principles. This is the position that I proposed in earlier work (e.g. Dupré 1993, ch. 2; see also Kitcher 1984). More recently, in the last section of Chapter 3, and in more detail in Chapter 4, I have taken a somewhat different tack, here approaching more closely to a genuine nominalism.

The philosophical problem over species derives from the fact that the concept of species is expected to serve two radically different functions. On the one hand it is assumed to be the fundamental level of biological classification, and on the other hand it is supposed to be a key theoretical term in the development of evolutionary theory. Those pursuing the first goal have often been led, like myself, to realize that there is no simple or unique criterion for assignment to this fundamental level, and have espoused some kind of pluralism. Most, however, being moved also by the second goal, have limited their pluralism to groups of organisms satisfying some criterion of evolutionary coherence (monophyly) (Ereshefsky 1992*a*; Mishler and Donoghue 1982). Others, like myself and Philip Kitcher (1984), have been more catholic in our pluralism, allowing that some species divisions may be grounded on morphological or ecological criteria. On the other hand those moved more by the second function of the species concept have been led to deny that species names are really proper classificatory terms at all. Here I have in mind the thesis associated with Michael Ghiselin (1974, 1997) and David Hull (1989) that species are not kinds at all but individuals, or more specifically, that species are chunks of the genealogical nexus.

The arguments for this thesis are in some respects compelling. Species, according to evolutionary theory, evolve and, perhaps, they are selected. To be the subjects or objects of such processes seems beyond the powers of an abstract entity such as a kind. But on the other hand the term species has been used for millennia to refer to the basic unit of classification, and it seems curious, to say the least, to claim that science has somehow discovered that species are not kinds at all. And moreover, as I argue in more detail in Chapter 4, biological organisms still need classification. The observations that evolutionary theory is much less well suited to providing classificatory principles for some kinds of organisms than it is for others, and that the distinctions offered by evolutionary principles are anyhow between individuals rather than between kinds, lead me to propose a radical solution to the species problem: there is no solution to the species problem as traditionally conceived. There remains a desideratum that organisms be classified in the most useful way possible. But it turns out that there is no unitary theoretical basis for this classification, but only a variety of different principles useful for different branches of the tree of life. Species are not evolutionary units but merely classificatory units, and classification should be pragmatic, pluralistic, and, broadly, nominalistic. Or so I argue.

In Chapter 5 I broaden the scope of the discussion, first by considering putative kinds in parts of biology other than taxonomy, and second by relating the nature of kinds to broader developments in our conception of science. It is certainly not the case that the members of particular species are the only candidates for biological natural kinds, even if they are the most traditional candidates. Within taxonomy itself it is now more common to suppose that species in general (the species category) form a natural kind, a thesis that fits well with the widely held view that particular species are not even kinds, but individuals. And there are other areas of biology. Two that I consider in this chapter are population ecology and genetics, though fairly cursory discussion is sufficient to point out that traditional conceptions of natural kinds do not apply to the central categories within these areas of science.

The traditional understanding of natural kinds is an important component of a broader view of science which, while subject to a good deal of criticism over recent decades, remains influential. This view sees science as discovering and formulating laws which, in turn, are the instruments through which explanation or prediction could be achieved. Natural kinds fit into such a picture by providing us with the appropriate subjects for fundamental scientific laws. As this picture has been criticized, an alternative has been developing, a picture that emphasizes not universal laws but local models. In this chapter I argue that the view of natural kinds I develop, as no more than the locally best way of addressing specific questions, is highly complementary to this general view of how science often or typically progresses. The chapter also elaborates the distinction between this view of natural kinds and more traditional, essentialist, views. The locally natural kinds I describe will vary greatly in character from one locality to another, and hence motivate a negative answer to the title question of this chapter, 'Is "Natural Kind" a Natural Kind Term?'

In Chapters 6 and 7 I turn from the general classification of organisms to the possibility of classifying the diverse members of a particular species, our own. There is a long tradition of attempting to understand human behaviour as a product of a universal human nature, a tradition of which sociobiology, and its recent successor evolutionary psychology, are the most prominent contemporary representatives. In Chapter 6 I present some criticisms of this view and suggest that the

kinds that are most promising for constructing behavioural generalizations are much smaller and more local than the entire human species,
kinds that I there refer to as cultural species. The analogy between
biological and cultural species should not be pushed too far. It does,
nevertheless, serve to emphasize the point that the processes that currently determine what behaviours are common, desirable, permissible,
or prohibited to people are of a quite different kind from those that
determined the patterns of behaviour of at least the vast majority of
our non-human ancestors.

A distinctive feature of evolutionary psychology, and one in terms
of which its practitioners attempt to differentiate themselves from their
somewhat discredited sociobiological precursors, is the emphasis on
the fact that our 'cognitive equipment', what it is that gives us the
capacity to engage in the complex behaviour characteristic of human
cultures, is a universal and evolved feature of our species. I do not,
of course, deny this. What I do deny is the fruitfulness of this observation for explaining the fine structure of human behaviour. (And perhaps also the claim that this fine structure is a superficial gloss on a
common coarse structure; I believe that human behaviour is fundamentally fine-structured.) This is an argument that I have developed
in considerable detail in later work (see Dupré 2001), and which can
only be introduced in this and the following chapter. My major
point here is rather to sketch the foundations for a complex and again
pluralistic view of the human sciences. A new level of complexity has
developed in recent history as traditional, partially isolated, cultures
have increasingly been merged into diverse multicultural societies.
This does not mean that the elements of traditional cultures have dis-
appeared. It does mean that these elements cannot easily be identified
with a distinct and separable group of individuals. In multicultural
societies people engage in a variety of more or less overlapping sub-
cultures. This suggests, in close parallel to what I take to be true of
the biological sciences, that scientific perspectives on human behaviour will need to be drawn from a large variety of inescapably partial
perspectives. This does not mean there is no truth in the human
sciences, only that there is no whole truth.

A corollary of what has just been said is that I see limited use for the
concept of normality as applied to human behaviour. In describing
the variation within a class of objects one might think of two different paradigms. Quality control inspectors apply one such model: there
is an ideal type of the objects they inspect and a range of deviations.

If the deviations are too great, the object is excluded. Variable features will often be normally distributed, and some distance from the mean will define the limit of acceptable deviation. An object that is in all respects close to the ideal type is naturally described as normal. The other paradigm admits no central, normal, or ideal type. Here we simply provide a system of categories to describe the variation or to locate particular objects within that range of variation. An extreme example would be the coordinate system for cartography. No one supposes the location (0,0) in a Cartesian coordinate system, or a location on the 0° line of longitude, is an especially normal place. In my view the classification of humans is much closer to the second paradigm than the first, and hence there is little content to the idea of a normal human.

Doubts about normal people are developed in Chapter 7 with reference to the concept of disability (and in greater detail in Dupré 1998). Certainly there are medical contexts in which it is appropriate to describe certain physiological states as normal or abnormal. Nevertheless, I believe that the scope of even this sense of normality is much narrower than is often supposed. It is tempting to understand human disability as simply a deviation from normal human physiology, but I think this temptation should generally be resisted. The issue turns on whether 'normal' abilities are best seen as intrinsic features of humans or relations between humans and their environments. Given that humans construct their environments, and construct them in ways that determine what physiological capacities are required for what abilities, I argue that the relational view of human capacities is the correct one.

I am even more sceptical about the usefulness of the related concept of human nature. Here, however, I need to confront directly a quite widely held opinion that not only is there such a thing as human nature, but proper attention to the legacy of Charles Darwin provides us with the key to its detailed exposition. This is a view that is also addressed in Chapter 7. There I discuss in more detail, and from a quite different perspective, the programme of 'evolutionary psychology' introduced in the previous chapter. Notoriously, debates in biology attempt to enlist Darwin on both sides, and I shall provide no exception to that rule. One of the major legacies of Darwin is to displace biological essentialism and replace it with a view of variation as fundamental to biological species. Although there are harmless enough interpretations of the expression 'human nature', the universalistic accounts developed by contemporary evolutionary psychologists belong, in my opinion, to a pre-Darwinian biology.

Darwinian evolution is, moreover, a dynamic, developing theory. It is reasonable to expect that our account of evolution in 100 years' time, say, will be far removed from our current understanding, and even further removed from Darwin's views. Evolutionary psychologists, on the other hand, take a frozen, one might say fundamentalist, view of the subject, and use it dogmatically to derive what they claim are the truths of human nature. Furthermore, I claim, certain current movements in evolutionary theory, notably the programme of 'developmental systems theory' and the recognition of the multi-level nature of selective processes, directly threaten many of the arguments of evolutionary psychologists. I do not undertake to defend these evolutionary ideas in any detail (though the second seems to me fairly solidly established) but rather to illustrate that Darwinian theory is nothing like the finished scientific product that evolutionary psychology typically assumes and requires it to be. Criticism of some of the detailed proposals of evolutionary psychology is provided in Chapter 9.

The most controversial of all classifications of humans is that provided by distinctions of sex or gender. It may (or may not) be that more damage has been done by the classification of people by race, but this is not a comparably contentious scientific issue. A decreasing minority of serious scientists thinks that race provides a biologically useful set of categories, but the scientific analysis of sexual distinction goes from strength to strength. There is of course a difference: no one would deny that there is an important biological basis to sexual difference, whereas race appears to have no interesting biological basis whatever. Still, the extent of this real biological basis to sexual difference is far from obvious. We can agree that there are two distinct and complementary roles in the physiology of reproduction. But the alleged ramifications of these roles into differentiation of behaviour is another matter. Feminist theorists have distinguished sex, as the biological distinction in reproductive physiology, from gender, the differentiation between men and women in social roles and behaviour. A powerful scientific tradition, however, aims to undermine this distinction and insists on the reduction of gender to no more than the behavioural ramifications of complementary reproductive roles. Currently evolutionary psychology, the successor to the sociobiology that achieved notoriety with E. O. Wilson's 1975 book of that name, is especially insistent that the diversity of gender roles emphasized by feminists is largely an illusion, and the same basic gender roles, manifestations of a universal human psychology, can be found in all human societies. Chapter 9 will confront some of the evolutionary psychological arguments directly. Chapter 8 is a more purely philosophical treatment of the assumptions underlying the kind of treatment of sexual difference exemplified by evolutionary psychology.

The philosophical doctrine that fits naturally with the sociobiological programmes is essentialism, perhaps the central critical target of this collection of essays, and it is in Chapter 8 that I provide the most detailed critique of this doctrine. Essentialism has meant many things throughout the history of philosophy. Here I have in mind a doctrine that relates directly to a conception of science. An essentialist view of science is one that sees science as involved in discovering the real kinds of things and in discovering what are the essential properties that make things members of those kinds.[1] This view is also directly challenged in Chapter 1 and somewhat tangentially in Chapter 5, to which the present chapter is complementary in obvious ways.

The more specific philosophical thesis of Chapter 8 is that essentialism involves a radical departure from an appropriately empiricist approach to science. There are countless quite legitimate ways of distinguishing entities in the world, and we should make no assumptions about the scope of such distinctions prior to real empirical investigation. The false assumption of an underlying essential nature to legitimate biological kinds was the error of Putnam's and Kripke's accounts of natural kind terms discussed in Chapter 1. In Chapter 8 the point is developed and elaborated with reference to the categories of sex and gender. There is no question but that sex is an important, indeed unavoidable, category for biology, or that gender is an equally necessary category for analysing human societies. Explicitly or implicitly this has encouraged the assumption that there must be some essential property that underlies sexual distinctions and explains the ramifications of sexual differences in physiology and behaviour. And since human gender distinctions coincide broadly with the sexual distinctions of male and female, this assumption mandates the attempt to assimilate human gender to biological sex.

---

[1] The most appropriate contrary to essentialism in this sense is the nominalism well described (and largely endorsed) by Ian Hacking (1999). In Hacking's sense I would surely count myself a nominalist, though the term also has connotations of anti-realism, which I would prefer to repudiate. I therefore prefer to call myself a promiscuous realist in the sense described in Ch. 1.

In this chapter, however, I argue that the essentialist perspective misleads about both biological sex and human gender. Although there are important theoretical reasons for the elaboration of the concept of sex, there is great variety in the manifestations of sexual differentiation. For a start there are organisms that are hermaphrodites or intermittently asexual, and species of protozoans that have considerable numbers of distinct mating types: the first one studied in detail has seven (see Nanney 1980; 1999: 98). But even with more familiar two-sexed species both the physiological basis of sexual difference and the behavioural ramifications are enormously various. There is no general account of sexual difference, no essence of male or female.

This observation already undermines a great part of the underlying motivation for the assumption that there must be some universal biological essence at some level deeper than superficial gender role diversity. And appealing to the general account of human diversity elaborated in Chapter 6, I argue that we should take human gender role diversity at face value. Human gender roles are adapted in different ways to the diverse human cultures of which they form crucial components. This is not to deny that gendered human brains develop as part of general human ontogeny and that this involves an interaction between internal biological causes and external cultural or environmental causes. But the empirical fact is that this interaction is capable of producing a great diversity of final psychological outcomes. There is no need, or room, for the assumption of any central, natural, normal, or biologically determined gender role.

In Chapter 9 I address directly the recent attempts by evolutionary psychologists (in my view thinly disguised sociobiologists) to illuminate gender-specific behaviour. This is the most polemical chapter in the volume. I review some of the explanations that evolutionary psychologists have offered for what they take to be sexually differentiated human behaviour, and claim that these have little or no merit. Central to their programme is a description of human life in the Stone Age, a description that is inevitably conjectural and that has been aptly referred to by Steven Rose as the Flintstones Theory: like the Flintstones, prehistoric humans are seen as leading lives essentially similar to suburban life in 1950s America. This conjectural prehistory has been supplemented by a good deal of empirical research on contemporary sexual behaviour and attitudes. Unfortunately much of this research is bedevilled by obvious methodological weaknesses. When predictions are confirmed, they are often of banal truisms, and when they are not confirmed, an indefinite range of ad hoc hypotheses is readily summoned to take care of the anomaly. No serious effort is made to rule out obvious alternative explanations in terms of cultural forces, forces the existence and importance of which I have tried to demonstrate in Chapter 6. I have criticized this work in more detail elsewhere, but the present chapter should give a good sense of the kinds of things evolutionary psychologists now say, and some of the reasons why they would better refrain from saying them. I conjecture that only bad philosophy supporting the idea that some account of the sort offered must be available can explain the willingness of scientists to produce such shoddy intellectual products.

Having discussed a variety of issues concerning both the classification of animals and the classification of humans, it seems appropriate to conclude with some discussion of the boundary between humans and animals. In some ways I suppose that this is one classificatory boundary that I am inclined to take more seriously than are many contemporary thinkers. I hold, that is, not only that human history is a very different kind of process from purely biological evolution, but also that it is one that produces comparable effects on the possibilities for human nature. And I have no doubt that there is a boundary between humans and other animals in the more obvious sense that there is no continuous range of extant intermediate cases between human and non-human animals. Nevertheless, I believe that the uniqueness of our species and its distinction from other species is often thought about in ways that reveal the malign influence of essentialism, and I try to demonstrate this in the last two chapters of this book. The two human excellences most often proposed as unique, perhaps essential, attributes of our species are thought and language, and I discuss these in turn in the last two chapters.

The idea that thought is a uniquely human attribute seems to me obscure, and on most interpretations highly implausible. The *locus classicus* for this idea is surely in the work of Descartes, and notoriously for Descartes thought (*cogitatio*) was understood in an astonishingly wide sense to include every kind of mental and even perceptual operation. Descartes does, at any rate, appear to have thought that no animal was capable of any kind of mental activity, and that no animal was conscious. Animals for Descartes were what many contemporary philosophers refer to as zombies, machines with no inner life. As the saying goes, the lights are on but nobody is home.

This extreme view seems uncommon nowadays, though not extinct (see e.g. Carruthers 1989).

My own view, expounded in Chapter 10, is that there are good reasons for applying most of our mental language to some animals, though no doubt the kinds of beliefs, desires, and suchlike that any non-humans are capable of entertaining are very different from most of our own. I claim, drawing on well-known arguments from the later Wittgenstein, that the common assumption that it is simply an open question whether other animals have any conscious experiences is a legacy of Cartesianism that should be abandoned. It is not an open question whether a chimpanzee with a badly broken leg showing obvious signs of acute discomfort might nevertheless really not be in pain. There are no further criteria, to use Wittgenstein's term, that the chimpanzee could exhibit that would show this; and the criteria for really being in pain are amply manifested. Similarly, I reject the intelligibility of the philosophical zombies just mentioned. Consciousness is not a wholly mysterious phenomenon totally disconnected from a self-sufficient system of stimuli, information-processing, and behaviour; if it were, we would have no way of knowing that *we* experienced it. 'An "inner process"', as Wittgenstein wrote, 'stands in need of outward criteria' (1953, §580). I don't think it should be controversial that many animals are conscious and have experiences of some sort. What kinds of experiences they have and what kinds of beliefs they are capable of entertaining, on the other hand, are much harder questions. But not, I maintain, unanswerable. Sophisticated students of animal behaviour with detailed knowledge of particular kinds of animals can develop well-grounded accounts of what their subjects are up to. Of course there is not, and nor could there be, any general account of animal minds. But there could be, and to some extent are, accounts of chimpanzee minds, cat minds, wombat minds, or porpoise minds (though these last, belonging to creatures whose lives are almost wholly alien to us, may be contingently inaccessible to us).

More difficult is the question of what capacities humans possess that are the prerogative only of creatures with a sophisticated language. This topic is a vast one, and would require another book, or several. It is, however, addressed tangentially in the final chapter of the book. There I look at some of the work that has been done on attempting to teach human, or human-like, languages to great apes. One theme of this chapter is that the arguments intended to show that these animals are incapable of any such language-learning whatever are unconvincing.

Typically, I suspect, they reflect an essentialist view of language that posits some condition necessary and sufficient for the possession of a true language, a condition which the animal subjects are assumed not to possess. I doubt whether there is any such essential condition, and am inclined to see continuity between the systems of communication of ourselves and of other higher animals.

Having said this, though I find these experiments fascinating, I am somewhat sceptical about what they show. I do think, however, that there are valuable morals to be drawn from them about language and about science. These experiments confront a dilemma that is reflected in two quite different approaches to their work. The earliest such experiments involved researchers who spent large amounts of relatively unstructured time interacting with their subjects, and claimed increasing and impressive abilities to communicate symbolically with them. From a scientific point of view such work was subject to a great deal of legitimate criticism. The claimed results were almost wholly 'anecdotal' and lacked any proper controls or replicability. It was difficult to convince a sceptical outsider that the animal was not simply being manipulated by the experimenter in ways familiar to generations of animal trainers.

The alternative approach involves a determined attempt rigorously to apply principles of good scientific method. This involved animals sitting in front of specially designed keyboards with minimal human interaction. The problem with this approach, it seems to me, is that though the animals acquired impressive skills at manipulating their environments using their keyboards, it is doubtful whether this could qualify as the use of language. Language is, after all, a social phenomenon, and if rigorous scientific method really requires, as some critics have maintained, that the experimental subjects be isolated from any social, especially human, contact while attempting to acquire their linguistic skills, then scientific method appears to preclude the investigation of the issue in question. The root of the difficulty is that the object of the experiments was to see whether apes can learn *human* languages. This surely amounts to the question whether an ape can be brought into a human linguistic community, something which experiments of the first kind, whatever their methodological deficiencies, seem much better suited to address.

But is also leads one to wonder whether the experiments are not, in the end, ill conceived. Surely what is interesting is whether there is an interesting and symbolically rich *ape* language, rather than whether apes can go some way towards learning our language. And this, surely, can only be discovered by detailed observations of communities of apes. Once again one suspects that an essentialist intuition is at work. If apes have linguistic abilities, then they have the ability to learn something essentially similar to a human language, and the object of the experiments is to see whether this is true. But if one supposes that the communicative skills of different species are related only by a diverse set of similarities, analogies, and differences, then teaching an ape a humanlike language is rather like teaching a dog to ride a bicycle. It may, depending on your tastes, be amusing. But the difficulty it presents to the dog should not lead us to conclude that dogs are deficient in locomotive skills.

In the end this last topic thus illustrates the most general theme of the essays in this volume. The world, and most especially the biological world, is a more diverse place than many scientific approaches to its investigation allow. There is a vast and variable array of similarities and dissimilarities between different biological kinds, and the attempt to apply general concepts across these kinds is always risky. In the end a great deal of biological investigation must be local and specific to particular groups or communities of organisms. And, at the risk of applying a general idea across disparate fields, I have argued that a similar situation applies to the study of humans. Perhaps there are some important truths that apply to all or almost all human societies, but this should not be assumed or even, perhaps, expected. Such universality does not follow from the undisputed fact that there is a great deal common to all humans in terms of physiology or genetics. Just as a knowledge of chemistry could give one no clue to the diversity of organisms that might evolve, so a knowledge of biology cannot enable us to predict the variety of cultures that may appear and, hence, the variety of human minds that might develop in those diverse cultures. On the whole the study of human history and human society must also, therefore, be irreducibly local. And, finally, the particularity of groups of organisms, as with the particularity of unique historical phases of human societies, is far too complex to be exhausted by any one theoretical approach. So not only must the biological and the human sciences be local, they must also be plural.

I
Kinds of Animals in
Everyday Life

1

Natural Kinds and Biological Taxa

The main topic of this paper is the theory of natural kinds that has been developed by Putnam (1975d, esp. 1975a, b, c) and Kripke (1972b; also 1972a). One area to which this analysis has seemed particularly appropriate is that of general terms naming biological organisms. My strategy will be to compare the requirements of this analysis with some actual biological facts and theories. It will appear that these diverge to an extent which, I will claim, is fatal to the theory. Toward the end of the paper I will also make some more constructive remarks about the nature of biological classification.

In the first section of the paper I will outline the theory in question, particularly as it has been developed by Putnam, and touch on some related historical and contemporary issues. In the second section I will assume the interpretation of biological taxonomy most favorable to Putnam's theory, and show that even this is often not as Putnam needs it to be. In the third section I will move to a more defensible account of biological taxonomy that renders the theory increasingly untenable. In the fourth section I will make some more construct-ive remarks about the relations between different ways of classifying organisms, and in the fifth and final section I will discuss the nature of species. The account I will offer, I believe, lends support to the contentions of earlier sections.

I

A good point of entry to the present issue is provided by Locke's theory of real and nominal essences. The distinction between real and nominal essence is, roughly, that between what accounts for the

I would like to thank Gordon Baker, who first suggested to me the philosophical interest of biological taxonomy, Nancy Cartwright and David Lewis, who made invaluable criticisms of earlier drafts of this paper, and most especially John Perry, who not only made many valuable criticisms of detail, but is also responsible for a much improved presentation of the entire paper.

properties characteristic of a particular kind ('the being of anything whereby it is what it is'; Locke 1975: III. iii. 15, p. 417), and the means whereby we distinguish things as belonging to that kind ('the abstract idea which the general, or sortal . . . name stands for'; Locke 1975: III. iii. 15, p. 417). For something like a triangle, which Locke took to be a wholly conceptual object, the real and nominal essences coincide. Since the properties of a triangle flow only from the way it is defined, contemplation of the latter could provide insight into the former. But one point of the distinction was to emphasize the futility of the scholastic, contemplative view of science. Contemplation of forms, nominal essences if anything, would be a source of knowledge of real substances only if nominal essences were also real essences. But they are not, so it is not. In the case of material things Locke, like his successors, thought that the real essence was some feature of the microscopic structure; i.e. that the microscopic structure was the real source of the phenomenal properties of a thing, and that microstructural similarities accounted for the homogeneity of macroscopic kinds. Of the practical value of this notion, on the other hand, Locke was sceptical. Regretting, famously, our lack of microscopic eyes, he doubted whether knowledge of real essences was possible, and also whether real essences, if they were discovered, would coincide with the nominal kinds we had previously distinguished. Thus he held that sorts of things were demarcated by nominal essences only. (Locke 1975: III. vi. 8, p. 443.) Subsequent scientific history has convinced some philosophers that Locke's scepticism was premature. Chemistry and physics have, since Locke's time, revealed a good deal about the microstructure of things, and antecedently distinguished classes of things have proved to share important structural properties.

The contemporary theory I want to discuss may now be crudely stated in Lockean terms as follows: (1) real essences demarcate natural kinds; (2) such natural kinds provide the extension of many terms in ordinary language. The theory does not attempt to conflate real and nominal essence. As we will see, Putnam has a theory of meaning that incorporates, and sharply distinguishes, both real and nominal essence. But it is the real essence that is supposed to determine the extension of the term. It is with the feasibility of this role that I will be mainly concerned.

Henceforward, I will use the term 'natural kind' to refer to a class of objects defined by common possession of some theoretically important property (generally, but not necessarily, microstructural).[1] The traditional view, to which Locke may be counted a subscriber, is that terms of ordinary language refer to kinds whose extension is determined by a nominal essence, and hence not to natural kinds;[2] and that science, on the other hand, attempts to discover those kinds that are demarcated by real essences. It is compatible with this view that in some cases real and nominal kinds will coincide. But this would be largely fortuitous. This position does not require that ordinary language is entirely independent of science, for several reasons. First, the explanation of our recognition of a kind might, in some cases, trace back to a theoretical feature that defined a natural kind. Second, terms that originate in scientific theory may become incorporated in ordinary language; we should certainly not suppose that these are separated by a sharp or impassable boundary. And third, it is widely accepted that even the most straightforwardly observational terms are to some extent 'theory-laden', though the exact extent of this is much debated. At any rate, the general picture is of science as a largely autonomous activity, in spite of subtle and pervasive interactions with the main body of language. It is one of the great attractions of Putnam's essentialism that it promises to provide much stronger links between science and ordinary language, since many terms of the latter are shown to refer to kinds demarcated by the former.

Putnam's theory resolves the meaning of a natural kind term into four components, referred to as a syntactic marker, a semantic marker, a stereotype, and an extension (Putnam 1975c: 269). To illustrate, the term 'elephant' might have as syntactic marker 'noun', as semantic marker 'animal', as stereotype 'large grey animal with flapping ears, a long nose, etc.', and an extension determined by the microstructural (or other theoretical) truth about elephants. It is with the last two of these, which are approximately equivalent to nominal and real essences (the stereotype being the nominal essence, stripped of its reference-fixing function), that I will be concerned.

---

[1] It should be noted that the expression 'natural kind' has sometimes been used in quite different ways. Quine, for instance, has used this expression in making the point that there are empirically discoverable distinctions in our 'subjective quality space' (Quine 1969). These kinds, however, depend on the particular nature of human observers, and not necessarily on objectively significant properties of the objects. They might, perhaps, better be referred to as 'innate nominal kinds'. Also, the discussion of species in the final section of the present paper could justify referring to species as 'natural kinds'; not, however, in the present sense.

[2] I use the term 'nominal essence' here very broadly to include definitions, criteria, clusters of symptoms, etc. I do not mean to imply that every kind requires an essential property.

The distinction between the stereotype and the extension is reflected in a distinction between mere competence in the use of a term, and (full) knowledge of the meaning of the term. The former requires only the first three components of meaning. In fact, the stereotype is explained as the set of features that must be known by any competent speaker of the language, regardless of whether it provides a good guide to the actual extension of the term. (Putnam 1975c: 25; 1975a: 204.) All this ignorant talk is facilitated by what Putnam describes as 'the division of linguistic labour' (1975c: 227–9). If, for any reason, it is important that items be assigned to the correct classes, it is necessary that there be experts familiar with the really essential properties of the kinds in question, and who are therefore able to perform this function. We generally take it on authority, for instance, whether something is made of gold. We may note, however, that we can never be sure even that the experts fully know the meaning of the term. For there is no guarantee that they have yet got right the real essence of the kind in question.

The central question raised by Putnam's analysis is how the nominal, or stereotypic, kinds of ordinary language are to be correlated with the natural kinds discovered by science. That is to say, granted that there are these real, empirically discoverable, natural kinds, how do we know which to assign to a particular term? Putnam answers this question by appealing to a previously unnoticed indexical component of meaning (1975c: 229–34). This consists in the reference, in using a natural kind term, to whatever natural kind paradigmatic instances of the extension of the term 'in our world' belong. Such a paradigm may be identified either ostensively, or operationally through the stereotype. Having identified the paradigmatic exemplar, the kind is then defined as consisting of all those individuals that bear an appropriate 'sameness relation' to this individual. This sameness relation is Putnam's exact equivalent of Locke's real essence. My fundamental objection to the theory as a theory of biological kinds is that no such sameness relations suitable for Putnam's theory can be found in it.

This concludes my exposition of Putnam's theory of natural kind terms. While I will argue that it is untenable, I should say now that there is much in it that I believe to be true. I am very ready to believe, in particular, that knowing the meaning of a term is something that admits of degrees, and that the higher degrees may only be achieved by experts. However, I do not think that experts can deliver on quite the task set for them by Putnam; and the task on which they can deliver, I think, is different in degree rather than in kind from what can be expected from a linguistically competent non-expert. Before attempting to substantiate these claims, I will conclude the present section with a brief consideration of the arguments that have been adduced by Putnam (and Kripke) in support of this kind of theory.

The general methodology that Putnam adopts is to consider counterfactual situations in which we encounter an item that is in some interesting respect novel, and then to decide (intuit?) whether we would apply a particular term to it. The relevant cases may be divided into two classes: first, those in which an object satisfies the stereotype, but for theoretical reasons is excluded from the extension, of a term; and second, those in which the theoretically important conditions are met, but part or all of the stereotype is not. I shall concentrate on cases of the first sort.

A favorite example of Putnam's is set in a place called 'Twin Earth'. This remarkable place is identical to Earth in every respect, except that what is there called 'water', a substance that plays exactly the role that water does on Earth and shares all the phenomenal properties of Earth water, turns out not to have the chemical composition H_2O, but to be some other complicated chemical substance, which may be called XYZ. Putnam's contention is that when we discovered this fact we would have to say that what they called 'water' was not water, since water is, necessarily, H_2O. Being H_2O is what constitutes the sameness relation for the natural kind, water. Since we have discovered that this is the appropriate sameness relation in our world, this fact has been incorporated in the very meaning of the term. The point I wish to emphasize here is a methodological one. If Putnam says 'XYZ is not water', and my intuition is that it would be (another kind of) water, how is such a dispute to be settled? Who knows what we ought to say in such a fantastic situation? Of course, the claim that XYZ would not be water must itself be intuitively plausible if it is to support, not merely illustrate, Putnam's theory.

Perhaps it will be helpful to notice that scientific history encompasses similar, if less extreme, cases. Consider, for instance, the first European botanist to study North American trees. When he arrived he might have been interested to discover that there were beech trees on that side of the Atlantic. More careful investigation would have told him that these beech trees differed from those he had previously encountered and in fact belonged to a distinct species.[3] Since the most striking difference between the two species was (perhaps) the size of the leaves, this discovery was commemorated in the distinction between *Fagus sylvatica* and the newly recognized *Fagus grandifolia*. In view of overwhelming similarities, there could have been little doubt about assigning these trees to the genus *Fagus* (= beech). Let us suppose that our botanist was also a linguist. If a native had asked him whether there were beech trees where he came from, what ought he to have said? My intuition, for whatever it is worth, is that he should have said that there were; though naturally if he were talking to a native *botanist*, he would go on to add that European beech trees belonged to a different species.

The purpose of this example is to suggest that plausible though some of Putnam's examples may be, they do admit of different interpretations. In the case of the water example it is also important to emphasize the great improbability of Putnam's hypothesis. All our scientific experience goes against the possibility of there being two substances that differed solely in having radically different molecular structures.[4] But this should not blind us to the fact that if we do take the possibility seriously, the best way of accommodating it might be to admit that there were natural kinds that encompassed such radical differences of structure. After all, it is surely just the absence of experiences like the one Putnam describes that makes it reasonable to attach to molecular structure at least most of the importance that Putnam ascribes to it.[5] Perhaps no one will be persuaded to take this case the way I have suggested. But I hope that I have at least said enough to motivate a closer look at how such issues are, and can be, treated in scientific practice.

---

[3] If it is objected that the concept of a species was very different when European botanists first reached America, I will make the modestly counterfactual assumption that America was first discovered in the 1970s.

[4] Indeed, if there is really *no* other difference, it is impossible to conceive of any ground there could be for postulating this difference. This may be seen as an example of what Schlesinger has called the Principle of Connectivity (Schlesinger 1963, ch. 3).

[5] Much of this importance may be attributed to the fact that Putnam is, or was, a reductionist. A classic statement of reductionist philosophy of science is Oppenheim and Putnam (1958).

II

Putnam's theory requires that there be kinds discriminated by science appropriate for providing the extensions of certain kinds of terms in ordinary language. A very encouraging source of examples for this thesis is available in biology, and it is these examples that I want to consider. The part of biology that is concerned with the classification of biological organisms is taxonomy. Within taxonomy, an organism is classified by assigning it to a hierarchical series of taxa, the narrowest of which is the species.[6] Thus a complete taxonomic theory could be displayed as a tree, the smallest branches of which would represent species. Rules would be required for assigning individual organisms to species, and an individual that belonged to a particular species would also belong to all higher taxa in a direct line from that species to the trunk of the tree. (In practice, an organism is classified by assigning it to successively narrower taxa. But as will emerge, this does not reflect the theoretical relations of successive taxonomic levels.) Let us assume what might be called 'taxonomic realism'. This is the view that there is one unambiguously correct taxonomic theory. At each taxonomic level there will be clear-cut and universally applicable criteria that generate an exhaustive partition of individuals into taxa. Each individual will then have the essential properties of all the taxa to which it belongs. We may even assume that the appropriate number of taxonomic levels to recognize is somehow implicit in the nature of the organisms. The claim that there are natural kinds in biology demarcated by real essences (and a fortiori Putnamian privileged sameness relations) would thus be entirely sustained. My first aim will be to show that even under these circumstances Putnam's theory faces serious difficulties of application.

The central difficulty I have in mind is that it is far from universally the case that the pre-analytic extension of a term of ordinary language corresponds to *any* recognized biological taxon. (Of course, I am not assuming that present biological theory includes the best possible taxonomy. But there can be no reason to anticipate a general trend towards coincidence with ordinary language distinctions.) In a sense this claim is not easy to substantiate, because the general terms in question are in fact extremely vague, and their application

[6] For some purposes divisions into subspecies or varieties are required. For the present discussion these can safely be ignored.

indeterminate. However, I think this indeterminacy can be seen to corroborate my thesis.

The richest source of illustrations for this difficulty is the vegetable kingdom, where specific differences tend to be much less clear than among animals, and considerable developmental plasticity is the rule. Any observant person who has explored the deserts of the south-west United States will have little difficulty distinguishing a prickly pear from a cholla. Yet taxonomically both these kinds of cacti belong to the same genus, *Opuntia*. Several species of this genus are certainly (to the ordinary man in the desert) prickly pears, and several are certainly chollas. Taxonomy does not recognize any important relation between *Opuntia polyacantha* and *Opuntia fragilis* (two species of prickly pear) that either does not share with *Opuntia bigelovia* (a species of cholla). Ordinary language does make such a distinction, and on the basis of perfectly intelligible and readily perceptible criteria. Thus the property of being a prickly pear is just not recognized in biology.

Or consider the lilies. Species which are commonly referred to as lilies occur in numerous genera of the lily family (Liliaceae). To take a few examples from the flora of the western United States again, the Lonely Lily belongs to the genus *Eremocrinum*, the Avalanche Lily to the genus *Erythronium*, the Adobe Lily to the genus *Fritillaria*, and the Desert Lily to the genus *Hesperocallis*. The White and Yellow Globe Lilies and the Sego Lily belong to the genus *Calochortus*; but this genus is shared with various species of Mariposa Tulips and the Elegant Cat's Ears (or Star Tulip). I would not want to undertake the task of describing the taxonomic extension of the English term 'lily'. However, it is fairly clearly well short of including the entire family. To include the onions and garlics (genus *Allium*, and, incidentally, another good example of the point of the previous paragraph) would surely amount to a debasement of the English term.[7]

It is not hard to find similar examples in the animal kingdom. The various species of chickadees and titmice share the same genus. Hawks probably comprise three of the four families in the order Falconiformae, though there are some questionable subfamilies. Whether a kite, an eagle, or a caracara is a hawk is another futile debate I will not attempt to initiate, though I feel sure that a vulture is not. Moths are another

---

[7] All the preceding examples may be found in Spellenberg (1969).

particularly interesting example. The order Lepidoptera includes the suborders Jugatae and Frenatae. It appears that all the Jugatae are moths. The Frenatae, on the other hand, are further subdivided into the Macrolepidoptera and the Microlepidoptera. The latter seem again to be all moths. But the former include not only some moths but also (all) skippers and butterflies.[8] In this case it does seem possible to give a plausible account of the taxonomic extension of the English word. The trouble is that the grouping so derived appears to be, from the taxonomic point of view, quite meaningless.

A rather desperate attempt might be made to save the theory from such examples, by going for the best available taxon and accepting some revisionary consequences for ordinary language. Thus one might claim that the extension of 'lily' was the whole family Liliaceae, or of moths the order Lepidoptera. We would just have to accept the fact that onions had turned out to be lilies, or butterflies moths. In defence of such claims, it could be pointed out that ordinary language has indeed come to accept such scientifically motivated changes as the rejection of the view that whales are fish in favour of the belief that they are mammals. But actually this example is by no means as clear-cut as is sometimes assumed. In the first place, 'mammal' is more a term of biological theory than of pre-scientific usage. One cannot recognize mammals at a glance, but must learn quite sophisticated criteria of mammalhood. 'Fish', by contrast, is certainly a pre-scientific category. What is more doubtful is whether it is genuinely a post-scientific category, for it is another term that lacks a tidy taxonomic correlate. I assume that the three chordate classes Chondrichthyes, Osteichthyes, and Agnatha would all equally be referred to as fish (unless sharks and lampreys are just as good non-fish as whales). But unless there is some deep scientific reason for lumping these classes together but excluding the class Mammalia, the claim that whales are not fish might be a debatable one. Perhaps 'fish' just means aquatic vertebrate, so that whales are both fish and mammals, and this well-worn example is just wrong. However, whales were never the most stereotypical fish, and it is easy to see the point of denying that they are fish at all: they do belong to a taxonomically respectable group most members of which do not remotely resemble fish. I see no parallel argument for the claim that butterflies are moths.

[8] For moths, see Borror and White (1970: 218 ff.).

The second difficulty for the application of Putnam's theory that occurs even against a background assumption of taxonomic realism concerns the hierarchical structure of taxonomy. Putnam's theory, it will be recalled, determines the extension of a natural kind term by means of a theoretical 'sameness relation' to a suitable exemplar. Suppose we want to discover the extension of the English word 'beetle'. A suitable exemplar will no doubt have to satisfy the condition that it be readily recognizable as a beetle by a linguistically competent layman; but probably this would not eliminate a very large proportion of the approximately 290,000 recognized species. Any particular exemplar will belong to one particular species. Given taxonomic realism, there will then be some sameness relation that it displays to other members of that species, some relation that applies within its particular genus, and so on up, not just to the relation that holds between all members of the order Coleoptera, which is approximately coextensional with the term 'beetle', but beyond, as far as the relation that holds between it and all animals but no plants. One may well wonder how the appropriate sameness relation is supposed to be selected from these numerous alternatives.

One kind of solution to this difficulty does suggest itself. If we collected a sufficiently large number of beetles, as different from one another as was consistent with the stereotype, we could try to find the narrowest sameness relation that held between every pair of our specimens. This methodology would, of course, force us to identify moths with the order Lepidoptera, and accept the consequence that butterflies were a kind of moth. It also seems to me that collecting the set of samples would involve attaching a lot of significance to the stereotype; if the stereotype were not a good guide to the real extension, it could hardly work. Rather than pursue this suggestion, however, I will now take a more critical look at taxonomic realism.

III

I have not meant to deny that very many general terms for living organisms do have a reasonably clear taxonomic correlate. But to investigate the extent of this correlation, it is first necessary to say something about the word 'ordinary'. For almost all species of birds and large vertebrates, for many flowering plants, and for some species of fish and insects, there is something (or sometimes a list of things) referred to as a common name. It is not obvious whether these should be thought of as part of ordinary language, or as part of a technical vocabulary. Certainly if competence in English does not require enough biological know-how to distinguish a beech from an elm (see Putnam 1975c: 226–7), then surely it cannot require an awareness even of the existence of the Solitary Pussytoes, the Flammulated Owl, or the Chinese Matrimony Vine. If such charming terms are assigned with their Latin equivalents to scientific taxonomy, and we restrict our attention to terms with which the layman can reasonably be expected to be familiar, then one thing we will find is that where there is a recognizable corresponding taxon, it is generally of higher level than the species.

For the case of large mammals, where human interest (and empathy) is at its highest, most familiar terms do refer to quite small groups of species; and common specific names are often widely known (as Blue Whale, Indian Elephant, or White-Tailed Deer). Most well-known names of trees refer quite neatly to genera, as, e.g., oak, beech, elm, willow, etc. (The various cedars, by contrast, are not closely related. It is reasonable to suppose that the term 'cedar' has more to do with a kind of timber than with a biological kind.) With birds the situation is highly varied. Ducks, wrens, and woodpeckers form families. Gulls and terns form subfamilies. Kingbirds and cuckoos correspond to genera, while owls and pigeons make up whole orders. The American Robin, finally, is a true species, though it is interesting that in Britain 'robin' refers to a quite different species, and in Australia, I am told, it refers to a genus of flycatchers. For insects, where the number of species is much greater, and the degree of human interest generally lower, the mapping is predictably coarser. Such things as hump-backed flies, pleasing fungus beetles, brush-footed butterflies, and darkling beetles make up whole families (the last-named, for instance, having some 1,400 known North American species). More familiar things, like beetles and bugs, refer to whole orders. (Must the competent speaker of (American) English know that a beetle is not a bug? Or is the word 'bug' ambiguous?)

The significance of the preceding point is that whereas there is an interesting case to be made for the reality of the species, there seems to be almost no case for taxonomic realism at any higher level of classification. Among biologists, 'lumpers' and 'splitters' do indeed dispute such questions as how many genera are to be distinguished within a family. Such disputes may be based on estimates of morphological or physiological similarity within groups of species, or on considerations of practical utility for field classification; they do not appear to involve deep theoretical interests, or to embody the assumption that such questions admit of true or false answers. (There is a possible claim that such distinctions reflect phylogenetic matters of fact, but I will postpone consideration of this suggestion.)

It will be recalled that Putnam's theory requires that there be some sameness relation between any two members of a natural kind. This might be called a 'privileged sameness relation' since it is not supposed to be just any relation that happens to demarcate the kind, but rather some discoverable relation that constitutes the real nature of that kind. But biological theory offers no reason to expect that any such privileged relations exist, since higher taxa are assumed to be arbitrarily distinguished and do not reflect the existence of real kinds. This claim will be reinforced in the final section of this paper, where I will argue that even for the case of species no privileged sameness relations exist. Since this is a rather more controversial question, however, I should emphasize that I do not think the argument against Putnam depends in any way on this question. For as I have indicated, a species is seldom a candidate for the extension of an ordinary language term.

To clarify my position on the relation between the species and higher taxa, I actually believe that the species is the only taxonomic level to which essential properties cannot be attributed. This is not meant as a paradox. It is merely that higher taxa, having no real existence, are defined in scientific vocabulary by nominal essences. Thus I would hold that such statements as that birds have feathers, mammals suckle their young, or spiders have eight legs are analytic.[9] But a nominal essence is not, in Putnam's sense, a privileged sameness relation. The reason that the same cannot be said of the species is that species, while lacking a real essence, do have a kind of real, objective existence. This will be explained in the final part of the paper. Meanwhile, there is one technical difficulty that may be raised against this account. Unless species have at least those properties that are essential to the

---

[9] This statement is considerably oversimplified. More strictly, we should say 'anything that has feathers is a bird', etc., to accommodate plucked birds, male mammals, and paraplegic spiders. However, 'anything that has eight legs is a spider' is neither analytic nor true. Further conditions that eliminate octopuses, crabs, etc., would be needed to construct a genuinely sufficient condition. Presumably the definition of a high-level taxon will typically be quite complex, and perhaps sometimes disjunctive.

higher taxa under which they are subsumed, it would appear that the relation between the species and the higher taxa cannot be subsumption. It might be sufficient to reply that a real essence is intended as a condition both necessary and sufficient for membership in a taxon, and this argument only shows that there are some necessary conditions. Since I am reluctant to admit that there are even strictly necessary conditions for species membership, I prefer a different line. One may assume that species are assigned to higher taxa *in toto*, because a sufficient majority of their members display the appropriate properties. (Thus, for instance, women who feed their infants from bottles would still count as mammals, since they belong to a species of mammals.) Then one would fail to assign an individual to a higher taxon only in case one failed to assign it to a species. This does not seem an objectionable failure.

IV

In this section I will make some more constructive suggestions about the relationship between the classifications of organisms in ordinary language (OLC) and in scientific taxonomy (TC). The natural way to contrast these classificatory schemes, it seems to me, is in terms of the different functions that they serve.

The functions of OLC, unsurprisingly enough, are overwhelmingly anthropocentric. A group of organisms may be distinguished in ordinary language for any of various reasons: because it is economically or sociologically important (Colorado beetles, silkworms, or Tsetse flies); because its members are intellectually intriguing (trapdoor spiders or porpoises); furry and empathetic (hamsters and Koala bears); or just very noticeable (tigers and giant redwoods). This list could no doubt be extended almost indefinitely, which merely reflects the immense variety of human interests. From this standpoint many apparent anomalies between OLC terms and TC terms are readily explicable. An example I mentioned earlier is illustrative here. It would be a severe culinary misfortune if no distinction were drawn between garlic and onions. But we have seen that this is not a distinction reflected in TC. Presumably there is no reason why taxonomy should pay special attention to the gastronomic properties of its subject matter.

A slightly more elaborate example is the following. The taxonomic classes birds and mammals are both part of ordinary language (though the latter less clearly). By contrast, the much larger class of angio-sperms (flowering plants) receives no such recognition. There is a very familiar term of ordinary language, 'tree', the extension of which undoubtedly includes oak trees and pine trees (though perhaps not their seedlings). The extension of the TC term 'angiosperm', on the other hand, includes daisies, cacti, and oak trees, but excludes pine trees. It is no surprise that such a grouping finds few uses outside biology; for most purposes it is much more relevant whether some-thing is a tree or not than whether its seeds develop in an ovary. This seems sufficient to explain why there is no taxonomic equivalent of 'tree' and no ordinary language equivalent of 'angiosperm'.

Where organisms are of little interest to non-specialists, they are typically coarsely discriminated in OLC. Thus it is that despite the vastly greater number of arthropod than vertebrate species, OLC dis-tinguishes many more kinds of the latter. The factors I mentioned before may all apply here. Vertebrates are more likely to be useful (nutritious), interesting (empathetic), furry (useful), noticeable (big), etc. Thus arthropod classifications in OLC typically cover enormous numbers of species. In fact, the useful distinctions tend to be more on the model of 'small red beetle' and 'large black beetle' than of specific identification. Still with this functionalist viewpoint in mind, we can also see that there may be other, specialized vocabularies that do not coincide with either TC or OLC. The vocabularies of the timber merchant, the furrier, or even the herbalist may involve subtle distinctions between types of organisms; there is no obligation that these distinctions coincide with those of the taxonomist. (Recall, for instance, my earlier suggestion about the term 'cedar'.)

TC, hopefully, avoids this anthropocentric viewpoint. The number of species names is here intended to reflect the number of species that exist. Nonetheless, even here there is an anthropomorphic aspect. For an adequate taxonomy must not only meet theoretical constraints, but should also be practicably usable. The strongest theoretical con-straints apply at the level of the species, for the obvious reason that this is the level with the greatest theoretical significance. Thus it has recently been recognized that a large number of groups that had been taken for species were in fact groups of very similar but distinct species.[10] There is no requirement that taxonomy must be easy.

---

[10] For a discussion of these so-called 'sibling species', see Mayr (1975, ch. 3).

A taxonomic system is not merely a list of species, but must also include a selection of features by which they are to be recognized. Such features may be called 'diagnostic'. If it were possible to discover some privileged sameness relation for species, then clearly this relation should be used as diagnostic for the species. In the final part of this paper I will consider and reject some candidates for such a relation. For now I will assume that the existence of a species consists in the general co-occurrence of a large number of characteristics. If this is right, then the selection of diagnostic features must be greatly under-determined, and hence, in a sense, arbitrary. Of course, there will be certain desiderata for such a choice, such as minimal developmental plasticity, or just ease of determination. In practice, a suitable feature or set of features is generally taken as providing a conclusive iden-tification. But this should not be taken as showing that the features selected are privileged. And indeed, a slight acquaintance with field biology suggests that even the best selected diagnostic features will occasionally fall foul of atypical specimens or obscure hybrids.

If the contrast I have suggested between species and higher taxa is well founded, it would be misleading to apply the term 'diagnostic feature' both to species and to higher taxa. For the latter, a better term would be 'defining feature'. As for the species, the fact that such a feature is not dictated by discoverable properties of the objects does not imply that there are no appropriate standards for selecting defin-ing features. Maximal evolutionary invariability is one desideratum that comes to mind.

The position I would like to advocate might be described as pro-miscuous realism.[11] The realism derives from the fact that there are many sameness relations that serve to distinguish classes of organ-isms in ways that are relevant to various concerns; the promiscuity derives from the fact that none of these relations is privileged. The class of trees, for example, is just as real as the class of angiosperms; it is just that we have different reasons for distinguishing them. It is true that in the case of species there is a largely, though not wholly, determinate range of classes that we are aiming to identify. The existence of species, I suggest, may be seen as consisting in the following fact. If it were possible to map individual organisms on a multidimensional quality space, we would find numerous clusters or

---

[11] I am grateful to John Perry for suggesting this term.

bumps. In some parts of biology these clusters will be almost entirely discrete. In other areas there will be a continuum of individuals between the peaks. It can then be seen as the business of taxonomy to identify these peaks. This picture also makes it easy to see why the deliverances of taxonomy need not provide the distinctions that are relevant for more specialized interests. As is demonstrated by the existence of sibling species, the properties that covary in a species and distinguish it from other similar species may be very subtle (at least subtle enough to have escaped biologists for a long time). When the classificatory problem is approached from a more restricted point of view, that is, with an interest only in a certain range of properties, many peaks will disappear, while others may be emphasized. As an example of the former, analysis of the vocalizations of frogs have revealed numerous sibling species. But this hardly need be a matter of concern to the gourmet unless there are also variations in the texture or flavour of frogs' legs. Again, the gourmet puts more emphasis on the distinction between garlic and onions than is implicit in taxonomy. Even within biology different interests call for the emphasis of different distinctions. Thus the primary unit of significance in ecology is not the species but the population.

V

In this concluding section I will defend the claim that privileged sameness relations cannot be found for the demarcation of the species.[12] At the same time, I hope to lend support to the positive characterization of the nature of species that was outlined in the previous section. While this account is certainly important for the general metaphysical position that I have just sketched, I should make clear that I do not think that the earlier claims about the relation between taxonomy and ordinary language depend upon the success of the present enterprise. These claims I take to be sufficiently established by the arguments adduced in Sections II and III of this paper.

I will now review and criticize three strategies that might be attempted for identifying privileged sameness relations between the

---

[12] The nature of species has received some discussion in the philosophical literature. For a view quite close to that presented here, see Ruse (1969). See also, e.g., Lehman (1967); Kitts and Kitts (1979).

members of a species. These strategies are based, respectively, on intrinsic properties of the individuals, on reproductive isolation of a group of individuals, and on evolutionary descent of a group of individuals. They will be considered in that order.

A traditional assumption that dates back at least to Aristotle is that organisms could be unambiguously sorted into discrete kinds on the basis of overt morphological characteristics. Since the theory of evolution undermined the belief in the fixity of species, this assumption has become increasingly untenable. It is now widely agreed that gross morphological properties are not sufficient for the unambiguous and exhaustive partition of individuals into species.[13] Crudely, this is because there is considerable intraspecific variation with respect to any such property, and the range of variation of a property within a species will often overlap the range of variation of the same property within other species.

At the same time it is still sometimes thought that a more covert, probably microstructural, property could be discovered that would be adequate for the unambiguous assignment of individuals to species.[14] More specifically, it may be thought that some description of the genetic material could capture a genuinely essential, or at least privileged, property.[15] It is assumed that the morphological and physiological properties are causally conditioned by interaction between the organism's genetic endowment and its environment. Thus it is imaginable that all members of a species do share the same genetic blueprint, or one with certain essential features, but that intraspecific differences are attributable to differences in environmental factors. But it is equally possible that there should be as much or more genetic variability as morphological variability. That is, intraspecific genetic variability may overlap interspecific variation as much as, or more than, morphological variability does. In fact, there are good reasons for supposing this to be the case.

There are various reasons why evolution should favour species with a high degree of genetic variability. In the first place, a reserve of genetic variety may enable the species to survive changing environmental conditions. A species, in other words, may be able to produce

---

[13] For a much more detailed discussion of this fact, see Mayr (1975, ch. 2).

[14] Kitts and Kitts (1979) argue that there must be such a property, while admitting that we do not yet know what it is.

[15] e.g. by Putnam; see Putnam (1975b: 141).

individuals suited to a variety of environmental situations. Second, it appears that heterozygous individuals (i.e. individuals with pairs of different genes at various loci) are often better adapted than homozygous individuals. (The more invariant the genetic material, of course, the less heterozygosity is possible.) A classic example of this is provided by sickle-cell anaemia. Only those individuals that are heterozygous with respect to this gene are able both to produce viable blood cells, and to exhibit a high resistance to a form of subtertian malaria which is prevalent in those areas where the gene in question is commonest. More generally, it is supposed that heterozygosity provides a way of increasing the diversity of the biochemical resources of an individual. Finally, it is believed that there are homeostatic developmental mechanisms whereby differing gene combinations approximate the production of the same phenotype.[16] This last point both accounts for the possibility that genetic variation might excede phenotypic variation, and also emphasizes why it would be mistaken to suppose that the genetic material was in any way privileged with respect to intraspecific homogeneity. Of course, there are other microstructural features that could be supposed to be especially favoured in this respect, such as the presence of particular proteins, lipids, or whatever. But there is no reason to expect that any such properties enjoy a privileged status with respect to variability.

Much importance is attached in theoretical biology to the notion of reproductive isolation. It is suggested that a species can be defined as a group of interbreeding individuals, reproductively isolated from all other individuals; this is often referred to as the 'biological' species concept, and may be considered a second candidate for providing a privileged relation between members of a species. Set against the desirability of genetic variation, there is a need for a species to maintain the integrity of a well-adapted gene pool. This requires insulation against the introgression of alien genes. Furthermore, it is generally supposed that the process of speciation is not completed until effective mechanisms have been established to prevent such introgression.[17] Thus there is a certain sense in which reproductive isolation is an essential property of the species: the species would not have come into existence if it had not, to a sufficient degree, acquired this property.

---

[16] See Mayr (1975: 133). Not surprisingly, sibling species are particularly likely to display strong developmental homeostasis.

[17] For a discussion of speciation, see Mayr (1975, chs. 15 and 16).

The important point here is that this is a property of the species, or gene pool, but only secondarily of the individuals that make up the species. An obvious way to make this point is to observe that bullocks or worker bees are not disbarred from species membership merely by virtue of being reproductively isolated from everything. This consideration is not, of itself, very convincing. Elaboration of the proposed criterion in terms of ancestral or other reproductive links to members of the interbreeding, but isolated, group might accommodate such cases. But deeper obstacles stand in the way of such a course.

Adequate reproductive isolation of a species does not require complete isolation of all its members. Hybridization occurs throughout the natural world, though more particularly among plants, fishes, and amphibians. (A recently publicized case of successful mating between two monkeys of different species has brought this fact to more general attention.) This need not lead to significant gene introgression. In some cases hybrid individuals are sterile. In more complex cases there may be a band of hybridization where the geographic ranges of two species meet. (A readily noticeable example occurs with primroses and cowslips.) In such cases the continued existence of the two species is made possible by the competitive superiority of each within its preferred range. Since this superiority will normally apply also over hybrids, the alien genes will not penetrate much beyond the area of overlap. Thus the suggestion that this criterion for species provides a privileged relation between its constituent individuals fails on two counts. First, there will be individuals that would not be assigned to any species on this criterion; and second, there will be reproductive links connecting individuals that certainly belong to different species. The latter point is reinforced by the fact that the ability to produce viable offspring is not transitive. There exist chains of species, any two adjacent members of which can produce viable offspring, but the terminal members of which are not able to interbreed. Finally, as has often been observed, this criterion is completely useless for asexual species, since it would imply that every asexual organism constituted an entire species.

The third, and final, proposal I will consider is one based directly on evolutionary history. The underlying idea is that it should be possible, in theory, to construct a family tree for all life on earth. It is then hoped that the classificatory taxonomic tree could converge on this phylogenetic tree. Hence any taxon will correspond to a historically real evolutionary process. This proposal has the considerable advantage that it appears to be equally applicable to the species and to higher taxa. Since it is certainly hoped that taxonomy and phylogeny should at least be mutually illuminating, this suggestion is in some sympathy with biological theory.

Two preliminary points should be made about this proposal. First, the essential or privileged property in question is highly extrinsic to the individuals to which it may be supposed to apply. Not only does it offer no hope of examining individuals and determining to which taxa they belong, but indeed nothing short of the entire evolutionary history of the organism will suffice for such a determination. The second point is related, though more speculative. The vast improbability that such a phylogenetic tree could ever be constructed does not seem wholly irrelevant. Much of the necessary theorizing depends upon traces left by organisms in the very distant past. The circumstances under which such traces remain are quite unusual, and the vast majority of organisms that ever existed probably left no trace whatever. (I suppose a determinist might be driven to deny this. But that, I think, is a problem for determinism.) Thus this is a case in which the underdetermination of a theory by all available evidence seems particularly probable. It is very likely that insisting on a phylogenetic criterion of taxon membership would make taxonomy literally impossible.

To assess the present hypothesis it is first necessary to explain how a taxonomic tree could also be interpreted as an evolutionary tree. This requires that something be said about speciation. *Qua* taxonomy, each taxon also includes all the lower taxa 'descended' from it. Thus the American Robin belongs simultaneously to the species *migratorius*, the genus *Turdus*, the family Turdidae, etc. The present suggestion interprets this as also embodying an evolutionary hypothesis. A species is composed of a number of populations that may be more or less differentiated from one another, both genetically and morphologically. When a population acquires some characteristic that isolates it genetically from the rest of the species, it is said to have achieved the status of a species. Thus the relevant evolutionary hypothesis would assert that at one time 'Turdidae' would have referred merely to a population of a larger species. Subsequently, this population would have achieved full species status, and still later divided into further species which now constitute the various genera in the family Turdidae. The particular genus *Turdus*, in turn, must have divided into further species, of which one is *migratorius*.

It remains to be seen whether this phylogenetic interpretation of the taxonomic tree can do anything to supply the taxon with a real essence, or privileged internal relation. Against the suggestion that evolutionary history could be essential to members of a taxon, one might deploy a Putnam–Kripke-type argument. If, say, a chicken began to lay perfectly ordinary walnuts which were planted and grew into walnut trees, I would not wish to refer to this result as the production of a grove of chickens. If accepted, this intuition shows that the right ancestry is not a sufficient condition for taxon membership. My intuition, moreover, is that the trees in question might prove to be genuine walnut trees, which is to deny that ancestry is even a necessary condition. However, having expressed suspicion of this style of argument, I do not want to rest any weight on this example of it. A more general argument is the following. Any sorting procedure that is based on ancestry presupposes that at some time in the past the ancestral organisms could have been subjected to some kind of sorting. One can imagine drawing up a phylogenetic tree and naming some branch of it; but the objective reality of the branch can be no greater than the objective reality of the grouping of organisms that constitutes the beginning of the branch. But I have claimed that, given all the organisms existing at a single time, there are no privileged properties or relations by means of which these can be sorted unambiguously and exhaustively into objectively significant classes. In short, the phylogenetic criterion must be parasitic on some other, synchronic, principle of taxonomy. It cannot generate privileged properties on its own.

As I have tried to stress, I do not mean to claim that species are unreal; only that they lack essential properties, and that their members cannot be distinguished by some privileged sameness relation. In fact, the existence of discrete species is one of the most striking and least disputable of biological data. If one examines the trees or birds in a particular area, it is apparent that these fall into a number of classes that differ from one another in numerous respects. But the essentialist conclusion that one might be tempted to draw from this fact is dissipated first by more careful study, which reveals that these distinguishing characteristics are by no means constant within the classes, and second by extending the scope of the investigation in both space and time, whereupon the limitations of both intraspecific similarity and interspecific difference will become increasingly apparent.

I think that a closer look at the nature of evolutionary history may help to clarify the position I am trying to present. Evolutionary history may indeed be perspicuously displayed in the form of a tree. Forks in the tree may be taken to represent the establishment of mechanisms for reproductive isolation between populations of a species, and unbroken lines to represent species that exist at a given time. If we interpret this tree as a graph whose abscissa is a measure of time, and whose ordinate represents some very complicated property measure,[18] then the lines can be taken as representing average members of the species. If we were to try and plot individual organisms on the same graph, presumably these would be distributed around the lines in a normal statistical way. Here we may note various complications of which the model should take account. First, the distance between the lines will be highly variable. In the case of higher mammals, for instance, the lines are generally well spaced. Thus there are few borderline cases for the application of such terms as 'man' or 'tiger'. If we take a number of sibling species of fruit flies, on the other hand, the lines may be very close together. This enables us to see how the number of species can be a determinate matter, whereas the assignment of individuals to species may be only partially determinate. For an analysis of the distribution of various properties of these fruit flies could reveal a series of sharply defined means, whereas some individuals might lie between the means for most properties. On the graphic model, some individuals will occupy positions intermediate between two (or in a multidimensional model, many) lines.

A further complication is that the process of speciation is not an instantaneous one. Also there are rare cases of two species merging. This, too, will be a gradual process. What this implies is that even when we draw a line through the taxonomic tree at a precise moment in time, the number of species in existence will not be wholly determinate. For where there are species in the process of dividing or merging, it will not be a determinable question whether there are two species or only one in existence at that time. This complication reveals a curious analogy with the preceding one. For again, there is a slightly different perspective from which essentially the same question does admit of a determinate answer. Retrospectively,

---

[18] Strictly, this must be envisaged as multidimensional. The present simplification is merely for expository ease. The present account offers at least a partial justification for the programme of numerical taxonomy, for which, see Sokal and Sneath (1963).

at least, it should be possible to say how many species existed during a certain period. For we can see whether species did in fact succeed in separating or merging during that period.

It is satisfying that this picture indicates a role for each of the proposed defining characteristics I have been discussing, in accounting for the existence of, and describing the nature of, the species. I believe that it also vindicates the reality of the species in a way that shows why none of these features can be sufficient to define the members of a species. And finally, as a consequence of this last point, it appears that even if terms of ordinary language did refer to species, Putnam's theory of natural kind terms still could not be applied to them.

2

Are Whales Fish?

The whale, the limpet, the tortoise, and the oyster . . . as men
have been willing to give them the name of fishes, it is wisest
for us to conform.

(Oliver Goldsmith (1728–1774); quoted in the
Oxford English Dictionary)

This chapter discusses the relationship between ordinary language
classification of biological organisms and scientific classification. In
a recent book (Dupré 1993) I argued at length for a pluralistic view
of biological classification. This thesis has two parts, a pluralism
about biological taxonomy and the insistence that folk taxonomies
are as legitimate, and can be interpreted as realistically, as scientific
taxonomies. A consequence of this, which I shall emphasize in this
chapter, is that there is no reason to expect folkbiological categories
to converge towards scientifically recognized kinds. Although there
are certainly cases in which this happens, the removal of whales from
the category of fish being a standard example which I shall consider
in some detail, I argue that such changes are, at least, philosophically
unmotivated.

The first thesis, that there is no unique and privileged biological classification, is somewhat less controversial. A similar position
has been endorsed by some philosophers (Kitcher 1984; Ereshefsky
1992*b*; but see Sober 1984*a*), and a related, if qualified, version has
been accepted by some systematists (Mishler and Donoghue 1982).
Although in many cases, when we look at organisms in a restricted
spatial and temporal location, biological distinctions are unambiguous,
even overdetermined, there is no reason to expect this to be the case
as we extend our perspective spatially and temporally. This follows
directly from our current understanding of the evolutionary process.
Given this indeterminacy of kinds in nature, we should anticipate
that different biological interests, for example, evolutionary versus
ecological or morphological, will often dictate different taxonomies.
Moreover, there is no reason to be concerned by the fact that biologists concerned with very different classes of organisms often appeal

to different classificatory principles. Whereas a classification aimed at reflecting phylogenetic history may have much to commend it for the study of birds or mammals, it may be neither feasible nor even useful for micro-organisms or, perhaps, flowering plants.[1]

The topic of this chapter, however, is the more controversial claim that folk classifications should be treated on a par with scientific classifications.[2] Scientific classifications, I argue, are driven by specific, if often purely epistemic, purposes, and there is nothing fundamentally distinguishing such purposes from the more mundane rationales underlying folk classifications. The opposite view has been promoted among philosophers in highly influential writings by Kripke (1972b) and, especially, Putnam (1975c). Putnam treated folk classifications as first approximations to scientific taxonomies and argued that it was inherent in our use of everyday natural kind terms that they be subject to refinement as scientific taxonomies were developed and improved. However, in presenting this argument, Putnam assumed erroneously that biological natural kind terms in ordinary language typically approximated scientific biological taxa. In fact this is not at all generally the case (see Chapter 1 in this volume).

Systematic divergence between scientific and folkbiological classification is readily documented for both Western and other indigenous peoples. It is sometimes suggested that similarities between scientific and indigenous taxonomies are so striking as to suggest some unique and objective set of facts by which both are determined. But closer examination shows this appearance to be illusory. First, most folk taxonomies are applied only to quite restricted areas of space and time, and as I have noted, from such a limited perspective morphological gaps between some kinds of organisms will be striking and unequivocal. Second, and more important, the large majority of folk taxonomy occurs at levels above that of the species, and higher-level classifications are not generally held to have much objective significance from a scientific point of view.[3] Only in restricted biological domains—birds,

---

[1] A detailed defence of taxonomic pluralism is provided in Ch. 3 in this volume.

[2] The equal legitimacy of scientific and folkbiological classifications is also argued by Atran (1998), who also argues that scientific classifications are, in an important sense, grounded on an antecedent folkbiology. Atran bases the autonomy of folkbiology, in part, on an innate cognitive structure, a thesis about which I am rather more sceptical, however.

[3] One classic study of folk classification argues that genera are the predominant level at which such classifications are made (Berlin *et al.* 1974). My impression is that the situation is much more varied for Western folk classifications, though this may very well reflect the more haphazard relations between contemporary urbanized people and the natural world.

mammals, some reptiles, and some flowering plants—do folk classifications sometimes extend down to the level of the species. Even in these domains, with the possible exception of mammals, species-specific terms are the exception rather than the rule. In our own culture many people can identify, more or less, a mouse, a sparrow, a frog, or a marigold. A Dusky Seaside Sparrow (*Ammospiza nigrescens*) or a Prickly Fetid Marigold (*Dyssodia acerosa*) is another matter. Indeed it is misleading to think of these latter as terms of general folkbiology. They are rather parts of the highly specialized vocabularies of birdwatchers and wildflower enthusiasts, folk who, despite the low esteem with which their activities are sometimes regarded by professional biologists, have a special concern to remain scientifically *au courant*. At any rate more familiar folkbiological terms correspond, at best, to higher-level categories such as the genus or family and frequently, as I shall illustrate, to no recognizable scientific category at all.

Let me add two parenthetical comments. First, the case of birdwatchers and other biological amateurs should remind us of the very important point that science does not have a monopoly on specialized vocabularies. It would be very interesting to compare the vocabularies of furriers, timber merchants, or herbalists with those of both biologists and ordinary folk. It seems likely, for example, that the rather disparate set of species that is referred to by the word 'cedar' is grouped together because of an interest in a particular kind of timber. Second, no doubt hunter-gatherer societies have more sophisticated resources for describing their biological environment than do typical Western suburbanites. I take it, though, that this amounts to no more than a matter of degree, and has no bearing on the contrast I want to draw between scientific and folk taxonomies. Hunter-gatherers, presumably, make fine distinctions among the plants and animals in their immediate environment because of a pressing interest in eating them or putting them to other practical uses. I take it that they would be less inclined than Westerners to defer to scientific revisions of their taxonomy unless persuaded of some practical benefit to be obtained from such a revision. In this regard, as I shall argue, I would take them to be revealing superior philosophical sophistication.

Returning to the main thread of the argument, I noted that most terms of folkbiology are not terms for species. Where folk taxa do refer to species, it is generally plausible to take the folk and biological terms as coextensive (though one might note that the two would respond differently to such scientific changes as the identification of sibling species). However, for very many cases in which the folk term refers to a higher level in the taxonomic hierarchy, there is not even a candidate referent from scientific biology. Two examples will be discussed in much more detail below. One very typical pattern is that a higher-level biological taxon will be divided into several folk taxa for reasons that have little or no biological salience. A few examples are frogs and toads, titmice and chickadees, hares and rabbits, or onions, garlics, leeks, and other more decorative lily relatives. The last example illustrates that often there are good non-biological reasons for such distinctions, in this case culinary. Other cases are more complicated. Petrels, for example, all of which are tubenoses (order Procellariiformes), include the Storm Petrels (family Hydrobatidae) and the Large Petrels, the latter comprising a few species in the family Procellariidae, the family that also includes all Fulmars and Shearwaters. And so on. At the opposite end of the spectrum we might finally note the various forms of *Brassica oleracea*—cabbage, kale, Brussels sprouts, cauliflower, broccoli, and the like, the last two even being assigned to the same subspecies, *botrytis*, which we certainly have ample reason to distinguish for more mundane, again culinary, purposes.

The general point then is that science and ordinary life provide disjoint taxonomies of the biological world. If, as Putnam suggests, we have linguistic intentions, implicit in our use of biological kind words, to refer to the kinds that science will eventually privilege as real, then these intentions will be radically subversive of ordinary language. Since there are often good and obvious reasons for our ordinary language distinctions, such subversion will be generally regrettable. Better then to resist these subversive intentions, and better yet to deny that anything so undesirable should be acknowledged as among our pre-theoretical linguistic intentions. Nevertheless, although our biological and folk taxonomies are disjoint, they often overlap locally and interact with one another in various interesting ways. I shall explore these interactions with reference to my main present example.

Why Whales should be Fish

In a paper some years ago (Chapter 1 in this volume) I claimed that, contrary to a widely held misapprehension, whales were perfectly respectable fish. I must confess, however, that I have since come to doubt this claim about whales. The reason for this, though, is not that I have come to see that the march of science was indeed irresistible. It is rather that I have come to doubt whether there is any sense in which such a well-entrenched tenet of folkbiology as that which excludes whales from the ranks of fish could be shown to be wrong. Thus I still think that folk once believed that whales were fish, and that they were duped into changing that belief for bad reasons. Before this unfortunate occurrence, whales indeed were fish. But I take it that now almost all educated people are quite confident that whales *aren't* fish. And this, I now think, is enough to show that they are right that whales are now no longer fish.

Let me begin by considering briefly what is a whale and what is a fish. We may, I think, be quite confident that these are terms of ordinary language. My 4-year-old son is a quite competent user of both, and I take it that he is not especially precocious in this regard. To begin with whales, I suppose that many, in the scientist spirit theorized by Putnam, would now try to identify whales with the order Cetacea. In support of my strong intuition that dolphins and porpoises are not whales, however, I am pleased to note that *Webster's* defines whale as 'any of the larger marine mammals of the order Cetacea, esp. as distinguished from the smaller dolphins'. Dolphins and porpoises do not form any significant subgroup of the order Cetacea. This order is, indeed, commonly divided into two sub-orders, namely the Toothed whales and the Baleen whales. Although dolphins and porpoises may constitute a specific lineage of the Toothed whales, proposing a dichotomy between this lineage, on the one hand, and all other lineages of toothed whales plus all the Baleen whales, on the other, makes no biological sense. Note also that dolphins are more closely related phylogenetically to Toothed whales than these latter are to Baleen whales. In sum, then, the category of whales is a biologically arbitrary one. In agreement with *Webster's* we may plausibly conclude that the reason dolphins and porpoises are not whales is much more prosaic: they aren't big enough. Very large size is central to the ordinary language concept of a whale. In fact the word 'whale' is often used as a metaphor for largeness or excess as, for instance, in 'a whale of a time'.

Similar remarks apply to the concept of fish. As the quotation at the beginning of this chapter suggests, the term is sometimes used very widely for any aquatic animal. This usage is suggested by the various aquatic animals, such as starfish, cuttlefish, crayfish, and generically, shellfish, whose names contain the suffix 'fish' but which most people would decisively exclude from the extension of that term. The *OED* in fact does offer the definition as 'any animal living exclusively in the water' for popular language but adds a 'scientific' usage which restricts the term to cold-blooded vertebrates with gills. *Webster's* begins with a version of this scientific definition and adds '(loosely) any of a variety of aquatic animals'. This scientific definition applies to members of the classes Chondrichthyes[4] (cartilaginous fishes), Osteichthyes (bony fishes), and Agnatha (jawless fishes). But there is little or no biological rationale for a category containing just these groups. The primitive and rather unpleasant species in the class Agnatha (lampreys, hagfishes, and slime eels) have little in common with a state-of-the-art carp or tuna. Sometimes, it is true, the agnaths are distinguished from the other two classes, only the latter being denominated 'true fish'. But in fact a shark and a salmon are barely more closely related to one another than either is to a lamprey.

Thus the notion that there is a 'scientific' usage of the word 'fish' is a decidedly suspect one.[5] The appeal in the definition to such technical matters as the possession of gills or cold-bloodedness seems rather a quasi-scientific rationalization of an extra-scientific linguistic intuition than the report of a genuinely scientific usage. Indeed, as is so commonly the case with attempts to define biological kinds, it is not even strictly true of all its intended referents. Some species of tuna maintain body temperatures as much as 20 degrees higher than their surroundings, and so should qualify as warm-blooded. And the lungfish *Protopterus* has been shown to get only 10 per cent of its oxygen from water through its reduced gills. If its gills were to disappear completely at a subsequent evolutionary stage, I doubt whether it would thereby cease to be a fish. Given, then, that neither 'whale' nor 'fish' is really a scientific term, the rationale for the dictum taught religiously to all our children that whales are not fish (and it is interesting that it is something that reliably requires to be taught) is more than a little unclear.

I have not attempted to do a historical study of this usage, but it is clear that whales have not always been non-fish. The whale and the sturgeon were once known as the royal fish. The great fish that

[4] This class is sometimes subdivided into two full classes, Selachii (sharks and rays) and Holocephali (rabbit-fishes, such as the grotesque *Chimaera*).
[5] I should note that this conclusion, as also the earlier one about whales, is quite independent of any special view about the nature of species.

swallowed Jonah is also referred to as a whale. And so on. Moreover, there are obvious reasons for including whales in the (non-scientific) category of fish. A dolphin is in many ways a similar beast to a tuna. Both are superbly adapted to swimming at high speeds, and consequently they show considerable analogous evolution in superficial morphology. Certainly they are a lot more superficially similar than either is to an eel, a sea horse, or a ratfish, or, for that matter, to a bat, a rabbit, or a sloth. Given, then, that whales were once seen as fish, and that neither 'whale' nor 'fish' is in any serious sense a scientific taxonomic term, it is interesting to speculate, at least, as to how they became non-fish.

Why Whales are not Fish

Despite the foregoing considerations, I have already conceded the obvious fact that whales are not fish. Why not? We might ask both why this fact is obvious, and why it came to be a fact at all. The first question is easy enough. Educated speakers will, I suspect, almost unanimously refuse to apply the word 'fish' to Blue whales, Killer whales, and similar creatures and, for that matter, to dolphins. Ultimately I suppose that this is the only relevant evidence, and that it is decisive. On the other hand, I also suspect that if pushed to rationalize this linguistic intuition, most people will be found to believe that scientists have found out what fish are, and what whales are, and that the latter are distinct from the former. Here, as I have argued, they would be mistaken. What a fish is is not the sort of thing a scientist (except, perhaps, a linguist) could find out.

Much more interesting, then, is the second question. Is there a good reason for teaching our children that whales are not fish? Even if these are not scientific categories, one might argue that some useful scientific knowledge is transmitted by using them this way. Whales are, after all, mammals, and no other mammals are much like fish. Being mammals ourselves, we tend to know quite a bit about this class of organisms, and we certainly learn a good deal about whales by knowing that they are mammals. But this argument is not compelling. The obvious rejoinder is that some mammals are fish. In fact, if we taught our children that whales were mammalian fish, they would both learn to apply general knowledge about mammals to whales (they bear live young and suckle them, are warm-blooded, etc.) and might also learn that 'fish', unlike 'mammal', was not a term for any coherent scientific grouping of organisms but a loose everyday term for (perhaps) any aquatic vertebrate. Indeed, the argument that because whales are mammals they cannot be fish seems to me to be a paradigm for the confusion between scientific and ordinary language biological kinds. (I shall return to this point in the next section.)

In some ways a more satisfactory answer to the present question might be a moral one. Certainly our sense of appropriate treatment of cetaceans is greatly influenced by the conviction that whales are not fish. Nowadays, if one offers someone the choice between a tuna sandwich or a whale steak, one will notice a strong tendency to choose the former and even some degree of outrage at being offered the latter. Even some people with moral objections to eating any mammals will sometimes consider it acceptable to eat fish. This perhaps points towards the greatest respect in which whales really do diverge from typical stereotypes of fish. We think of fish as being fairly stupid creatures and of whales, even by mammalian standards, as being decidedly bright. Although no doubt the emphasis on intelligence in deciding how bad it is to kill and eat something has an anthropocentric aspect—intelligence being of course what we take to be so special about ourselves—there is surely something plausible to the idea. Killing a highly intelligent animal will perhaps cause great grief to its relatives and schoolfriends, and perhaps even will put an end to a more intrinsically valuable life. But certainly this point has at most rhetorical force. There is no reason why some kinds of fish should not be much more intelligent than others. Certainly there are such differences among mammals (between a human or whale, say, and a shrew) and perhaps there are comparable distinctions in intelligence between different kinds of non-cetacean fish. Moreover, I take it that whales ceased to be fish long before anyone was much troubled by eating them—indeed in many countries it appears that they are still not much troubled by doing so. So certainly this is not the historical explanation for the change.

For want of a better explanation, I conclude that the exclusion of whales from the category of fish developed as a response to greater scientific knowledge of the nature of whales, which was, as I have argued, a somewhat misguided response. No doubt such infiltrations of ordinary language by science are common enough. So I now turn to a more general consideration of the relation between scientific and ordinary language terms for biological kinds.

Science and Ordinary Language

The general point of the discussion so far is to insist that ordinary language and science provide largely independent and often disjoint ways of classifying the biological world. Where ordinary language biological kinds are distinct in extension from any coherent scientific kind, the attempt to revise them in accord with supposedly scientific discoveries can promote only confusion. Moreover, as I have argued elsewhere (Chapter 1 in this volume; Dupré 1993), I take it that ordinary language classifications are typically quite as well motivated, and the kinds to which they refer may be just as objectively real, as biological classifications. It is just that they are differently motivated. Thus onions may constitute a perfectly real class of organisms, but not one that biologists have found any good reason to classify together. There are presumably objective facts about organisms of certain species that make them suitable for flavouring stews though I must admit that I cannot confidently vouch for the fact that all the species referred to as onions meet this condition. Similarly with fish, though the rationale for this much broader category is perhaps a lot vaguer. On the other hand, it is clear that to call these systems of classification *wholly* independent would be an exaggeration. For better or worse, as the principal example in this chapter illustrates, the development of scientific thinking sometimes has an impact on ordinary language usage.

One important aspect of this interaction is the importation of relatively pure scientific categories from scientific taxonomy into ordinary language. I have already suggested that this is the way that many species terms should be understood. Here the amateur naturalists who are the primary users of such terms have an explicit interest in distinguishing those species recognized by professional biologists, and ordinary language terms are treated as synonyms for Latin binomials. Perhaps of greater interest are the constructions of ordinary language terms to acknowledge higher-level taxonomic distinctions developed by biologists. Several examples have occurred in the foregoing discussion. One of the most familiar examples is 'mammal'. This, surely, is a term imported into common usage from scientific taxonomy. Less familiar, but similar, examples are the jawless fish, cartilaginous fish, and bony fish also mentioned above. The point of these terms is to be able to make biologically significant distinctions in somewhat more user-friendly ways than by learning terms such as Osteichthyes or Chondrichthyes. Whereas the latter serve purposes such as international standardization important for biologists, the former are both easier to remember and more informative for less technical scientific or quasi-scientific discourse. I take it, however, that 'is a member of the class Agnatha' and 'is a jawless fish' are perfect synonyms, with scientific practice determining the reference. (Being a jawless fish does not mean being a fish and being jawless, or it would apply to all those bony fishes on the fishmonger's slab that have had their heads cut off.)

I want to claim, however, that these imported terms from scientific discourse should be understood quite differently from more familiar and well-entrenched ordinary language terms, and that the failure to make this distinction indicates a significant confusion common to many philosophers and lexicographers. Whereas the definition of mammal as, say, 'warm-blooded, hairy vertebrate with a four-chambered heart and which nourishes its young with milk from maternal mammary glands' is entirely appropriate, the definition of fish as 'cold-blooded, aquatic vertebrate with gills, and (usually) scales, fins, etc.' is much more questionable. These definitions look very much alike. But, I have suggested, the terms to which they apply are of quite different kinds. Thus, while it is quite appropriate to say that it is a scientifically attested fact that all mammals have four-chambered hearts, it seems to me something like a category mistake to say that it is a scientific fact that all fish have gills—not because they might not but because science has nothing to say about all fish.

This last remark perhaps gets to the heart of the present problem. It is, I think, not widely accepted that there are any matters of much importance about which science has nothing to say. And certainly the question, 'What kind does this organism belong to?' will strike most people as paradigmatically the kind of question about which science must be the only authoritative arbiter (e.g. see again Putnam 1975c). Ironically, perhaps, the idea of an authoritative and unique answer to such a question really assumes some version of essentialism, the idea that some fundamental, essential property of a thing makes it the kind of thing it is, and essentialism was central to the Aristotelian and Scholastic views of knowledge against which modern science developed in large part as a critical reaction. Essentialism in biology, more specifically, was delivered its death blow by the triumph of Darwinism, and the consequent recognition that an organism might belong to a quite different kind from its ancestors, and that variation rather than uniformity was the norm for a biological kind. In a biology premissed on variation and change, there is no reason to expect any unique answer to questions about how organisms should be grouped together, and a fortiori, there is no reason to expect science to provide such answers.

I do not want to deny the possibility that there might be scientifically motivated revisions of ordinary language biological kinds for which the revision might be sufficiently justified by some biological knowledge that it would somehow convey, though I am not convinced that there are such. Certainly there are many cases in which we change our views as to what higher-level, scientifically defined, kinds familiar ordinary language kinds belong. But as in the case of whales and fish, whereas it is certainly of interest to recognize that whales are mammals, this fact has no bearing that I can see on whether they are fish. The majority of cases that come to mind are of just this kind. So, for instance, it is sometimes said that marsupial mice are not really mice. The point, of course, is that Australian mice are very distantly related to mice in most of the rest of the world. For the slightly more technical this reflects the divide between marsupial and placental mammals, and I take it that these kinds, unlike mice, certainly, are pure scientific imports. (Indeed I take it that while the notion of a marsupial is a fairly familiar one, many fewer people can correctly contrast it with a placental. Poststructuralists might note this as a case of an unmarked category—like male or white—which is invisible because assumed by default.) Since 'mouse', on the contrary, is an ancient and well-entrenched term of ordinary language, the denial that marsupial mice are real mice is again based on an illegitimate interplay between categories of quite different kinds.

It may be that excessive scientism, whether among lexicographers, high school science teachers, or just regular folk, will continue to favour a continuing convergence between scientific and ordinary language taxonomies. If this is the case, it does not reflect a gradual Peircean convergence on some objective reality but, rather, the hegemonic power of one, sometimes imperialistic, method of knowledge production. Perhaps for most folk in the West such imperialism is relatively harmless; for most of us urban and suburban folk what we call organisms doesn't matter very much. And even for folk more intimately connected with nature, I do not suppose that adopting Western scientific modes of classification would be very damaging to their dealings with natural objects. Probably in those cases in which they would be damaging, folk will refuse to adopt them.

Nevertheless, there are reasons for resisting, or at least pointing out, this imperialism. With regard to its effects on Western culture, the main such reason is simply to resist the excesses of scientism. The achievements and successes of science are amply evident, but it is also important that human culture has aims and projects that are distinct from and incommensurable with those of science, and science does not hold the answer to every question of human interest. Quasi-scientific doctrines such as that whales have been discovered not to be fish help to obscure this point. With regard to our view of non-Western biological classification, this moral may be more important. Especially in relation to cultures with more regular and direct interaction with nature, we would do well to explore thoroughly the basis and function of such classifications before criticizing them for their non-convergence on our own scientific categories. Once again, the perception of the value of Western science will hardly be enhanced by insisting on unsubstantiated claims to insight where this is not to be had.

Promiscuous Realism

I conclude this chapter with some brief remarks of a more purely philosophical character. Views about classification tend to assume a strictly dichotomous form. On the one hand, there is a strongly realist view. Things belong to natural kinds, and it is part of the agenda of science to discover what these kinds are. Such a view is naturally, though not invariably, associated with essentialism: discovery of a natural kind involves discovering the essential property that qualifies a thing as a member of the kind. On the other hand, there are strong forms of constructivism. Kinds, on such views, are not natural but reflect grids imposed upon nature by humans. Nature is seen as amorphous until structure is imposed on it by us. It is generally assumed that such a view makes classification in some sense arbitrary. Perhaps there are human ends that favour one scheme of classification over another, but at any rate nature is equally amenable to any such scheme. I have advocated an intermediate view, however, which I have called promiscuous realism (Chapter 1 in this volume; Dupré 1993).

According to promiscuous realism, there are many, perhaps very many, possible ways of classifying naturally occurring objects that reflect real divisions among the objects. But not just any arbitrary classification will reflect real such divisions. Thus my position is realist, in that I insist that there is something in nature that legitimates a good set of classifications; in fact I see no harm in claiming that good classifications reflect natural kinds so long as the conception of natural kind in question is sharply separated from any connection with essentialism, and so long as it is recognized that a thing may belong to many different natural kinds. But the position also recognizes an ineliminable role for human classifiers in selecting a particular classification scheme from among the many licensed by natural similarities and differences. This selection will of course depend crucially on the purposes for which the classification is intended. Indeed it might be that relative to a particular purpose there is a uniquely best scheme of classification. What is excluded is that there should be a uniquely best system of classification for all purposes or, which comes to the same thing, independent of any particular purpose. The underlying philosophical view I have criticized as scientistic in preceding parts of this chapter is the view that science can be expected to provide just such a goal-independent set of classifications. The reason I reject this idea is not, as is generally supposed by constructivists, that nature provides no objective divisions between kinds but rather that it provides far too many. This is the promiscuity in the thesis of promiscuous realism.

Biology provides the central illustration of this view. Both scientific and extra-scientific discourses provide a variety of different ways of classifying organisms. I argue that the classifications favoured by science are distinguished from the rest not by their superior objectivity but simply by the specific, though various, goals that characterize scientific investigation of nature. If biological science could lay claim to a uniquely correct scheme of classification, it might be hard to resist the claim that this was in some sense *the* uniquely correct scheme for all purposes.[6] It is therefore important to the defence

---

[6] I say it *might* be hard. Actually I am not at all sure it might not nevertheless be possible and desirable to resist this claim. It is widely assumed that chemistry has provided just such a uniquely best scheme for classifying kinds of stuff by reference to chemical structure. But there are purposes for which it would be quite wrong—legally and economically, for example—to classify diamonds and graphite together, despite both being forms of elemental carbon. In the other direction, an example that has been discussed in the philosophical literature is that of jade, which comes in two quite distinct chemical forms, jadeite and nephrite, which are not relevantly different in terms of their properties as semiprecious stones.

of promiscuous realism that there are good grounds for denying the existence of any such unique classificatory scheme even for biological science (see above, and for more detail, Dupré 1993, ch. 2; Chapter 3 in this volume). If different theoretical or practical scientific goals dictate different principles of classification, it is difficult to see why the various schemes of biological classification in ordinary language or in technical but non-scientific domains should not be accorded equal objective legitimacy.

Thus, finally, as well as misrepresenting in detail the scope of biological discovery, the denial that whales are fish propagates, in my view, bad philosophy. It reflects the assumption that inclusion of one kind within another can only reflect a positioning of the subordinate kind within a unique hierarchy of kinds, the hierarchy gradually being disclosed by biological science. But in fact there are many such partially overlapping and intersecting hierarchies. This situation would be usefully highlighted by the much more perspicuous claim that whales (and dolphins and porpoises) were mammalian fish.

Regrettably, I have had to admit that whales are not fish, for the sufficient reason that almost everyone in our culture, ignoring the partially excellent advice of Goldsmith, agrees not to call them so. Most folk assume, I take it, that biological kinds either fully include one another or are wholly disjoint. And thus since certainly not all mammals are fish (and vice versa), none can be. It would be futile and ridiculous for me to attempt a campaign for the reinstatement of whales into the realm of fish. Nevertheless, the recognition that there is no good reason for their having been excluded from this category would be salutary.

II
Kinds of Animals
in Biological Science

3

On the Impossibility of a Monistic Account of Species

... if we can once and for all lay the bogey of the existence of true relationship and realize that there are, not one, but many kinds of relationship—genealogical relationship, morphological relationship, cytological relationship, and so on—we shall release ourselves from the bondage of the absolute in taxonomy and gain enormously in flexibility and adaptability in taxonomic practice.

(J. S. L. Gilmour, 'The Development of
Taxonomic Theory since 1851')

By the classification of any series of objects, is meant the actual, or ideal, arrangement of those which are like and the separation of those which are unlike; the purpose of this arrangement being to facilitate the operations of the mind in clearly conceiving and retaining in the memory, the characters of the objects in question. Thus there may be as many classifications of any series of natural, or of other, bodies, as they have properties or relations to one another, or to other things; or, again, as there are modes in which they may be regarded by the mind.

(T. H. Huxley, *Introduction to the
Classification of Animals*)

Most of the philosophical difficulties that surround the concept of species can be traced to a failure to assimilate fully the Darwinian revolution. It is widely recognized that Darwin's theory of evolution rendered untenable the classical essentialist conception of species. Perfectly sharp discontinuities between unchanging natural kinds could no longer be expected. The conception of sorting organisms into species as a fundamentally classificatory exercise has nevertheless survived. Indeed, the concept of a species traditionally has been the

I would like to thank Rob Wilson and Chris Horvath for valuable comments on a draft of this essay.

paradigmatic unit of classification. Classification is centrally concerned with imposing conceptual order on diverse phenomena. Darwin's theory, as the title of his most famous work indicates, is about the origins of diversity, though, so it is no surprise that the dominant task in post-Darwinian taxonomy has been to connect classificatory systems to the received, Darwinian, account of the origin of diversity. Attractive though this task undoubtedly is, it has proved unsuccessful. The patterns of diversity that evolution has produced have turned out to be enormously diverse, and in many cases the units of evolutionary analysis have proved quite unsuitable for the basic classificatory aims of taxonomy. Or so I argue.

Why do we classify organisms? A natural and ancient explanation —expressed clearly by, for example, Locke (1975: III. v. 9) and Mill (1862)—is that we do so to facilitate the recording and communication of information. If I tell you some animal is a fox, I immediately convey a body of information about its physiology, habits, and so on. The more you know about animals or mammals or foxes, the more information about that particular animal I convey. If organisms came in sharply distinguished natural kinds, internally homogeneous and reliably distinguishable from the members of any other kind, then the identification of such kinds would be the unequivocal aim of taxonomy. A classificatory system that recognized such natural kinds would be unequivocally the best suited to the organization and dissemination of biological information. But this is just what Darwin has shown us we cannot expect (see e.g. Hull 1965). In a domain of entities characterized, in part, by continuous gradation of properties and varyingly sharp and frequent discontinuities, matters are much less clear. It is this fact about the biological world that makes attractive the idea of taxonomic pluralism—the thesis that there is no uniquely correct or natural way of classifying organisms and that a variety of classificatory schemes will be best suited to the various theoretical and practical purposes of biology.

Many biologists and philosophers appear to think that pluralistic accounts of species will lead us to Babel (see e.g. Ghiselin 1997: 117–21). Biologists, they suppose, will be unable to communicate with one another if they are working with different species concepts. In this paper, I argue that species pluralism is nevertheless unavoidable. However, I also defend a kind of minimal monism: to serve the traditional epistemic goals of classification, it is desirable to have one general set of classificatory concepts. However, this general taxonomy will need to be pragmatic and pluralistic in its theoretical bases. For specialized biological purposes, such as the mapping of evolutionary history, it may often be necessary to adopt specialized classificatory systems. My monism is merely semantic: I suggest it would be best to reserve the term *species*—which is, as I have noted, the traditional philosophical term for classificatory concepts—for the base-level categories of this general, pragmatic, taxonomy. Such an anti-theoretical concept of species will discourage the conspicuously unsuccessful and controversial efforts to find a solution to the 'species problem', and leave it to working biologists to determine the extent to which they require specialized classificatory schemes for their particular theoretical projects.

Monists, needless to say, disagree about which actual species concept biologists should accept. The cheapest way to buy monism might be with a radically nominalistic phenetic concept, as conceived by numerical taxonomists (Sneath and Sokal 1973). If biological classification could be conceived as merely an exercise in recording degrees of objective similarity, then some particular degree of similarity could be defined as appropriate to the species category. But few people now think this can be done. Philosophically, attempts to construe a notion of objective similarity founder on the fact that indefinitely many aspects of difference and of similarity can be discovered between any two objects. Some account of what makes a property biologically interesting is indispensable: there can be no classification wholly innocent of theoretical contamination. Without wishing to deny that phenetic approaches to classification have provided both theoretical insights and practical benefits, I restrict my attention in this essay to more theoretically laden routes to species monism. My conclusions, however, leave entirely open the possibility that a version of pheneticism, modified by some account of what kinds of properties might be most theoretically interesting, may be appropriate for important domains of biology. The classification of bacteria is a likely example (see e.g. Floodgate 1962 and further discussion below).

In the section 'Troubles with Monism', I trace some of the difficulties that have emerged in attempting to provide monistic accounts of taxonomy motivated by central theses about the evolutionary origins of diversity. I thereby hope to substantiate my claim that as more has been learned about the diversity of the evolutionary process, the hopes of grounding therein a uniform account of taxonomy in general, or even the species category in particular, have receded. In the final section, I outline my more constructive proposal for responding to this situation.

Troubles with Monism

The potential conflict between two main goals of classification has long been recognized. The first and most traditional goal is to facilitate the communication of information or to organize the vast quantities of detailed biological information. From this point of view, a taxonomy should be constructed so that knowing the taxon to which an organism belongs should tell us as much as possible about the properties of that organism. This goal must, of course, be qualified by pragmatic considerations. Indefinite subdivision of classifications can provide, theoretically, ever more detailed information about the individuals classified: assignment to a subspecies or a geographical race will presumably give more information than mere assignment to a species. As the basal taxonomic unit, the species should be defined, therefore, to classify organisms at a level at which the gains from finer classification would be outweighed by the costs of learning or transmitting a more complicated set of categories. If organisms varied continuously with no sharp discontinuities, this balancing of costs and benefits would present a largely indeterminate problem. By happy chance for many kinds of organisms there appear to be sharp discontinuities at a relatively fine classificatory level that are much sharper than any discontinuities at any lower level. To the extent that this is the case, the selection of the appropriate level for assignment of organisms to species appears unproblematic.

In recent years, this goal of organizing biological information has been emphasized much less than a second, that of mapping the currents of the evolutionary process. A recent anthology of biological and philosophical essays on the nature of species carried the title *The Units of Evolution* and the subtitle *Essays on the Nature of Species* (Ereshefsky 1992b). Though the idea that, by definition, species should be the units of evolution is not uncontroversial, it is widely held. What is a unit of evolution? Evolutionary change is not change in the properties of any individual organism, but change over time in the distribution of properties within some set of organisms. (We need not worry here whether these properties are conceived as genetic or phenotypic.) A unit of evolution is the set of organisms in which changes in the distribution of properties constitute a coherent evolutionary process.

Because an evolutionary change is one with the potential to be maintained in future organisms, it is easy to see that the temporal dimension of a unit of evolution must be defined by relations of ancestry. As long as we are concerned with biological evolution in which properties are transmitted genetically (and ignore some complexities of gene exchange in bacteria), then evolution will be constrained within sets of organisms defined temporally by parent–offspring relations. We must then consider what determines the synchronic extent of a unit of evolution. A natural and attractive idea is that a species should include all and only those organisms with actual or potential reproductive links to one another. This condition would determine the set of organisms among whose descendants a genetic change in any member of the set might possibly be transmitted. To the extent that the biological world is characterized by impenetrable barriers to genetic exchange, then there will be distinct channels down which evolutionary changes can flow. The sets of organisms flowing down these channels, then, will be the units of evolution.

Here, of course, is the great appeal of the so-called *biological species concept* (BSC)—until recently the dominant conception of the nature of species. According to this view, a species is conceived as a group of organisms with actual or potential reproductive links to one another and reproductively isolated from all other organisms. Recalling for a moment my brief discussion of classification as mere ordering of information, one might also suppose that the sharp discontinuities that (sometimes) determine the optimal level for making base-level discriminations should correspond to lines of reproductive isolation. The flow of evolutionary change down reproductively isolated channels, after all, should be expected to lead to ever-growing morphological distinctness. Thus, the goals of representing the evolutionary process and of optimally ordering biological phenomena would turn out to coincide after all.

Unfortunately, however, the biological world proves much messier than this picture reveals. Certainly, there are cases in which species can be identified with discoverable lowest-level sharp discontinuities marked by reproductive barriers. But such cases are far from universal, and the appealing picture drawn thus far has a range of important complications to which I now turn.

Asexual Species

A familiar objection to the BSC is that it has nothing to say about asexual species. A fully asexual organism is reproductively isolated from everything except its direct ancestors and descendants. The leading proponent of the BSC, Ernst Mayr, has concluded that there are, strictly speaking, no species of asexual organisms (Mayr 1987). But asexual species still require classification, and indeed some asexual species are more sharply distinguishable from related species than are some sexual species. Moreover, asexual organisms evolved just as surely as did sexual species. Thus, whichever view we take of the fundamental goal of assigning organisms to species, the exclusion of asexual organisms should lead us to see the BSC as at best one species concept among two or several concepts necessary for encompassing biological reality. A more radical attempt to save the BSC is suggested by David Hull (1989): in asexual organisms, the species are simply organism lineages —that is, an organism and its descendants (p. 107).[1] I take it that although Hull's proposal is attractive theoretically, it will divorce the identification of species in these cases from any practical utility in classification. It should also be noted that even this radical move may not work to give the biological species concept universal applicability. In bacteria, although reproduction is asexual, various mechanisms are known by which bacteria exchange genetic material. The pattern of relationships between bacteria is thus netlike, or reticulated, rather than treelike.[2] Although I suppose that one might hope to identify a new species as originating at each node in the net, such an identification would imply the existence of countless species, many lasting only a few minutes or even seconds. The impracticality of this idea suggests that we would be better abandoning the idea of applying

---

[1] It is not entirely clear how to make this idea precise. Obviously, not every organism founds a lineage, unless every organism is to constitute a species. A natural idea is that every organism in any way genetically distinct from its parent should found a new lineage. Given, however, the possibility of the same point mutation occurring more than once, it could turn out that a set of genetically identical organisms might constitute two or more distinct species.

The proposal also leads to the surprising conclusion that the vast majority of species are asexual. As Hull notes, this conclusion may mitigate the well-known difficulty in explaining the origin of sex by showing that sexual reproduction is a much rarer phenomenon than is often supposed (1989: 109). I should also mention that Hull's proposal is made in connection with the thesis that species are individuals, and is thus not necessarily an explicit defence of the BSC.

[2] It appears that the same is probably true for some kinds of flowering plants (see Niklas 1997: 74 ff.).

the BSC, or indeed any evolutionarily based species concept, to bacteria. Many bacterial taxonomists (see Nanney 1999) indeed seem to have this inclination.[3]

Gene Flow beyond Sharp Discontinuities

A second familiar difficulty with the biological species concept is that apparently well-distinguished species frequently do, in fact, exchange genetic material. The classic illustration is American oaks (see Van Valen 1976). Various species of oaks appear to have coexisted in the same areas for millions of years while exchanging significant amounts of genetic material through hybridization. Ghiselin (1987) is quite happy to conclude that these oaks form a large and highly diversified species. Two responses should be offered to this conclusion. First, and most obviously, the need to make such a move illustrates the divergence between this kind of theoretically driven taxonomy and the pragmatic goal of providing a maximally informative ordering of nature. This divergence may not much bother the theoretically inclined, but it does illustrate one of the ways in which we cannot both have our cake and eat it in the way indicated in the most optimistic explication of the BSC.[4] Second, such examples throw serious doubt on the central motivation for the BSC, which is that genetic isolation is a necessary condition for a group of organisms to form a coherent unit of evolution. The example shows that different species of oaks have remained coherent and distinct vehicles of evolutionary change and continuity for long periods of time. Ghiselin's conclusion looks like

[3] This claim is perhaps less true now than it was twenty years ago. An influential evolutionary classification of bacteria was proposed by Woese (1987); see also Pace (1997). On the other hand, Gyllenberg et al. (1997) aim explicitly to produce a classification that is optimal from an information-theoretic perspective, a goal that there is no reason to suppose would be met by any imaginable phylogenetic scheme. See also Vandamme et al. (1996) for a related proposal. It is clear, at any rate, that any possible phylogenetic classification of bacteria, if it is to be of any practical use, must define taxa with great clonal diversity. Gordon (1997), for instance, reports that the genotypic diversity of *Escherichia coli* populations in feral mice was an increasing function of the age of the mouse, indicating the development of distinct clones during the lifetime of the mouse. I assume one would not want to think of this development as speciation, but given this clonal diversity, it is difficult to see how any useful taxonomy could avoid being arbitrary from a phylogenetic perspective. The situation is still worse in view of the partially reticulate phylogeny consequent on genetic exchange between bacteria.

[4] An extreme statement of this optimistic view can be found in Ruse (1987: 237): 'There are different ways of breaking organisms into groups and they coincide! The genetic species is the morphological species is the reproductively isolated species is the group with common ancestors.'

nothing more than an epicycle serving solely to protect the BSC from its empirical inadequacy.

The Absence of Sharp Discontinuities

In some groups of plants and of micro-organisms, and very probably in other kinds of organisms, there is considerable variation, but no apparent sharp discontinuities. It is even tempting to suggest that within certain plant genera there are no species. A good example would be the genus *Rubus*, blackberries and their relatives. Because *Rubus* lacks sharp differentiation between types, but admits great variation within the genus as a whole, it seems unlikely that there could be any consensus on its subdivision into species.[5] If we assume that this lack of sharp differentiations is due, in part, to gene flow, the option is again open to call *Rubus* a single and highly polymorphic species. Though less objectionable than in the case where there are sharply distinguished types, as with oaks, this move again separates theory-driven taxonomy from the business of imposing useful order on biological diversity.

Lack of Gene Flow within Sharply Differentiated Species

A somewhat less familiar point is that a considerable amount of research has shown that often there is surprisingly little genetic flow within well-differentiated species (Ehrlich and Raven 1969), most obviously in the case of species that consist of numbers of geographically isolated populations, but that nevertheless show little or no sign of evolutionary divergence. Even within geographically continuous populations, however, it appears that genetic interchange is often extremely local. This kind of situation puts great weight on the idea of *potential* genetic flow in defending the BSC. If populations are separated by a distance well beyond the physical powers of an organism to traverse, should their case nevertheless be considered one of potential reproduction,

---

[5] On *R. fruticosus*, the common blackberry, Bentham and Hooker (1926: 139) wrote: 'It varies considerably. The consequence has been an excessive multiplication of supposed species . . . although scarcely any two writers will be found to agree on the characters and limits to be assigned to them.' The same 'species' is described by Schauer (1982: 346) as 'aggregate, variable with very numerous microspecies'. More optimistically, *The Oxford Book of Wildflowers* (Nicholson *et al.* 1960) states that 'There are several hundred species and hybrids in the *Rubus* group, and only an expert can identify all of them.'

on the grounds that if, *per impossibile*, the organisms were to find one another, they would be interfertile? The alternative, paralleling Ghiselin's line on oaks, would be to insist that such apparent species consisted of numbers of sibling species, differentiated solely by their spatial separation. Again, one is led to wonder what the point of either manoeuvre would be. Clearly, to the extent that species retain their integrity despite the absence of genetic exchange, it must be concluded that something other than gene interchange explains the coherence of the species. Contenders for this role in cases like either of the kinds just considered include the influence of a common selective regime and phyletic or developmental inertia. I might finally note that although I do not know whether any systematic attempt has been made to estimate the extent of gene flow in the genus *Rubus*, in the likely event that the flow is quite spatially limited, the claim that the whole complex group with its virtually worldwide distribution can be seen as reproductively connected is tenuous to say the least.

The conclusion I want to draw at this point is that the BSC will frequently lead us to distinguish species in ways quite far removed from traditional Linnaean classification and far removed from the optimal organization of taxonomic information. Moreover, the theoretical motivation for the BSC seems seriously deficient. The sorts of criticisms I have been enumerating above have led, however, to a decline in the extent to which the BSC is now accepted, and this decline has been accompanied by increasing interest in a rather different approach to evolutionarily centred taxonomy that can be broadly classified under the heading of the *phylogenetic species concept* (PSC). (The definite article preceding the term should not be taken too seriously here, as there are several versions of the general idea.)

The central idea of all versions of the PSC is that species—and, in fact, higher taxa as well—should be monophyletic. That is, all the members of a species or of a higher taxon should be descended from a common set of ancestors. An appropriate set of ancestors is one that constitutes a new branch of the phylogenetic tree. Such a group is known as a *stem species*. The important distinction between versions of PSC is whether a taxon is merely required to contain *only* descendants of a particular stem species or whether it is required to contain *all* and only such descendants. The latter position is definitive of cladism, whereas the former, generally described as evolutionary taxonomy, requires some further criterion for deciding which are acceptable subsets of descendants.[6] Two issues arise in explicating a more detailed account of the PSC. First, what constitutes the division of a lineage into two distinct lineages and hence qualifies a group as a stem species? Second, what constitutes a lineage and its descendants as a species (or, indeed, as any other taxonomic rank)?

The traditional answer to the first question is that a lineage has divided when two components of it are reproductively isolated from one another, but the difficulties raised in connection with the BSC suggest that this answer is inadequate. Examples such as oaks suggest that reproductive isolation is not necessary for the division of a lineage, and worries about the lack of gene flow within apparently well-defined species suggest that it is not sufficient either. An illuminating diagnosis of the difficulty here is provided by Templeton (1989), who distinguishes *genetic* exchangeability, the familiar ability to exchange genetic material between organisms, and *demographic* exchangeability, which exists between two organisms to the extent that they share the same fundamental niche (p. 170). The problem with asexual taxa and with a variety of taxa for which gene exchange is limited is that the boundaries defined by demographic exchangeability are broader than those defined by genetic exchangeability. Conversely, for cases in which well-defined species persist despite gene exchange, the boundaries defined by genetic exchangeability are broader than those defined by demographic exchangeability (p. 178).

In the light of these considerations, Templeton proposes the *cohesion species concept* (CSC). It is not entirely clear how this concept should be interpreted. In the conclusion of his paper, he writes that species should be defined as 'the most inclusive group of organisms having the potential for genetic and/or demographic exchangeability' (p. 181). If we assume that the connective 'and/or' should be interpreted as inclusive disjunction, this definition would suggest that the 'syngameon' of oaks—that is, the set of distinct but hybridizing species —should be treated as a species. But it is clear from earlier discussion that such an application is not what Templeton intends. Earlier, he defines the CSC as 'the most inclusive population of individuals having the potential for phenotypic cohesion through intrinsic cohesion mechanisms' (p. 168). A central and convincing motivation for this definition is the claim that a range of such mechanisms promotes

---

[6] See Sober (1992) for a very clear exposition of this distinction. Although the debate here is a fundamental one, it is not of central concern to my essay.

phenotypic cohesion, of which genetic exchange and genetic isolation are only two. Equally important are genetic drift (cohesion
through common descent), natural selection, and various ecological,
developmental, and historical constraints. The basic task, according
to Templeton, is to 'identify those mechanisms that help maintain a
group as an evolutionary lineage' (p. 169).

What, then, is an evolutionary lineage? The significance of the
conflicting criteria of genetic and demographic exchangeability is
that they show it to be impossible to define the lineage in terms
of any unitary theoretical criterion. Rather, lineages must first be
identified as cohesive groups through which evolutionary changes flow,
and only then can we ask what mechanisms promote this cohesion,
and to what extent the identified groups exhibit genetic or demographic
exchangeability. Presumably, this initial identification of lineages
must be implemented by investigation of patterns of phenotypic
innovation and descent over time. With the abandonment of any
general account of speciation or any unitary account of the coherence
of the species, it appears that species will be no more than whatever
groups can be clearly distinguished from related or similar groups.
This approach may seem theoretically unsatisfying, but to the extent
that it reflects the fact that there are a variety of mechanisms of speciation and a variety of mechanisms whereby the coherence of the
species is maintained, it would also seem to be the best concept we
can hope for.

This conclusion makes pressing the second question distinguished
above: How do we assign taxonomic rank, especially species rank, to
a particular lineage or set of lineages? A prima facie advantage of the
BSC is that it provides a clear solution to this problem: a species is
the smallest group of individuals reproductively connected (or at least
potentially connected) one to another and reproductively isolated from
all other individuals. The difficulty is that this definition would leave
one with species ranging from huge and diverse syngameons to clonal
strains with a handful of individuals. Apart from the theoretical
difficulties discussed above, any connection between the theoretical
account of a species and a practically useful classification would
surely be severed.

The question that must be faced, then, is whether from the PSC
point of view the idea that the species is the basal taxonomic unit—
where taxonomy is conceived as providing a practically useful
classification—can be maintained. Abandoning the BSC will take care

of species that look unsatisfactorily large by allowing a variety of cohesion mechanisms apart from reproductive isolation, but it will tend to imply the presence of disturbingly small species. Frequently there are clearly distinguishable groups of organisms—subspecies, varieties, geographical races—below the species level. There is no reason to suppose that these groups are not monophyletic and no reason to suppose that they are not, at least for the moment, evolving independently. There is no doubt that such groups are often clearly distinguishable, and indeed for many purposes classification at this level is the most important. Stebbins (1987: 198) notes, for instance, that foresters are often more concerned with geographic races than species and indeed can be hampered in their work by the confusing attachment of the same specific name to trees with quite distinct ecological properties and requirements. A judge at a dog show is not much concerned with the criteria that identify something as *Canis familiaris*. Such groups may go extinct, they may merge with other subgroups in the species, or they may be destined to evolve independently into full-blown species or higher taxa. Their evolutionary significance is thus unknown and unknowable. The same, of course, could be said of groups recognized as full species, though the second alternative (merging with other groups) may be rare.

The Case for Pluralism

An evolutionarily based taxonomy appears to be faced at this point with only two possible options. The first is to consider species as by definition the smallest units of evolution. Leaving aside the insurmountable difficulty of detecting such units in many cases, my argument so far has been that this option will provide a fundamental classification that is often much too fine to be useful for many of the purposes for which taxonomies have traditionally been used.[7] Mishler and Donoghue (1982) suggest that this proposal is also conceptually confused. They argue that 'there are many evolutionary, genealogical units within a given lineage . . . which may be temporally and spatially overlapping' (p. 498). They suggest, therefore, that it is an error to suppose that there is any such thing as a unique basal evolutionary

---

[7] See Davis (1978: 334–8) for a discussion of some of the difficulties in subspecific classification of angiosperms.

unit and that the particular evolutionary unit one needs to distinguish will depend on the kind of enquiry with which one is engaged. If there is no unique basal unit, then there is no privileged unit and, from an evolutionary point of view, no theoretical reason to pick out any particular group as the species. Mishler and Donoghue therefore propose the second option, to 'Apply species names at about the same level as we have in the past, and decouple the basal taxonomic unit from notions of "basic" evolutionary units' (p. 497). This process involves seeing species on a par with genera and higher taxa— that is, as ultimately arbitrary levels of organization, chosen on a variety of pragmatic grounds.[8]

Although Mishler and Donoghue see the species as an ultimately arbitrary ranking criterion, they do maintain a version of the PSC and, hence, do not see it as arbitrary from the point of view of grouping. In fact, they endorse the strong, cladistic concept of monophyly as a condition on a group constituting a species (or, for that matter, a taxon at any other level). Their 'pluralism', however, entails that 'comparative biologists must not make inferences from a species name without consulting the systematic literature to see what patterns of variation the name purports to represent' (p. 500). But given this degree of pluralism, and the rejection of the attempt to equate the basal taxonomic unit with any purportedly fundamental evolutionary unit, one may reasonably wonder why it is desirable to insist nevertheless on the requirement of monophyly. I suspect that part of the motivation for this requirement is the idea that there must be *some* answer to the question what a species *really* is. It was once, no doubt, reasonable to suppose that evolution had produced real, discrete species at approximately the classificatory level of the familiar Linnaean species. Perhaps this supposition was an almost inevitable consequence of the transition from an essentialist, creationist view of nature to an evolutionary view. Acceptance of evolutionary theory would require that it more or less serve to explain biological phenomena as theretofore understood. Nevertheless, a further century of development of the evolutionary perspective has given us a radically different picture of biological diversity. The sharpness of differentiation between kinds and the processes by which such differentiation is produced and

---

[8] For further elaboration, see Mishler and Brandon (1987). For more general arguments against any fundamental distinction between species and higher taxa, see Ereshefsky (1991, 1999) and Mishler (1999).

maintained have proved to be highly diverse. There is no reason to suppose that evolution has provided any objectively discoverable and uniquely privileged classification of the biological world.

Why, then, should we continue to insist that evolution should provide a necessary condition, namely monophyly, on any adequate biological taxon? I can think of only three possible answers. First, it might be held that a better understanding of evolution is so overwhelmingly the most important biological task that any taxonomy should be directed at improving this understanding. Second, it might be thought that an evolutionarily based taxonomy, despite its problems, would provide the best available taxonomy, or at least a perfectly adequate taxonomy, for any biological project even far removed from evolutionary concerns. Or third—and this, I suspect, is the most influential motivation—it may be held on general methodological grounds that a central concept such as the species must be provided with a unitary definition. This third motivation might be grounded either in a general commitment to unification as a scientific desideratum or on the fear that failure to provide a unified account of the species category will lead to massive confusion as biologists attempt to communicate with one another. I argue, however, that none of these proposed justifications of the demand for monophyly stand up to much critical scrutiny.

The first answer can be quickly dismissed. Even as distinguished an evolutionist as Ernst Mayr (1961) has emphasized the distinction between *evolutionary* and *functional* biology, the former being concerned with questions about ultimate causation (how did a trait come to exist?), the latter with questions of proximate causation (how does the trait develop or function in particular individuals?). Following Kitcher (1984), I prefer to distinguish these types of questions as historical and structural. It is clear that questions about the ontogeny of the human eye, say, or about the processes by which it provides the individual with information about the environment, have little to do with questions about how humans came to have the kinds of eyes they have. Of course, just noting this fact doesn't show that we need a taxonomy based specifically on structural aspects of organisms, but it does remind us that there is more to biology than evolution. A particularly salient domain, about which I say a bit more below, is ecology.

We should turn, then, to the second, and more promising, line of thought. The fact that a great variety of kinds of investigation takes place within biology certainly does not show that one scheme of classification, based on phylogenetic methods, might not be adequate to all these purposes. To some degree, it should be acknowledged that this question is purely empirical: only the progress of biological enquiry can determine whether different overlapping schemes of classification may be needed. This point needs to be stated carefully. There is no doubt at all that interesting structural or physiological properties cross-cut any possible phylogenetically based classification. An investigation into the mechanics of flight, for instance, will have relevance to and may appeal to a group of organisms that includes most (but not all) birds, bats, and a large and miscellaneous set of insects. In general, convergent evolution and the acquisition or loss of traits within any sizeable monophyletic group make it clear that no perfect coincidence between monophyletic groupings and the extension of physiologically interesting traits can be anticipated. Whether this calls for a distinct, non-phylogenetic system of classification is less clear. To pursue the example given, there is no particular reason why the student of flight should attach any particular significance to the miscellaneous group of organisms that fly.

Ecology, on the other hand, raises more difficult issues. Ecology, it may be said, is the microstructure of evolution. Nevertheless, it is not obvious that evolutionarily based taxa will be ideal or even well suited to ecological investigations. Certainly, there are categories—predator, parasite, or even flying predator—that are of central importance to ecological theory and that include phyletically very diverse organisms. There is no reason why phyletically diverse sets of organisms might not be homogeneous (for example as fully substitutable prey) from the perspective of an ecological model. On the other hand, such concepts may reasonably be treated as applying to a higher level of generality than the classification of particular organisms. At a more applied level of ecology, however, some kind of taxonomic scheme must be applied to the particular organisms in a particular ecosystem. Ecology will often be concerned with the trajectory of a population without addressing competition between different subgroups within that population. It may, that is, abstract from distinctions within a population, perhaps corresponding to distinct lineages, which could be fundamental in understanding the longer-term evolutionary trajectory of the population. Groups of sibling species may prove ecologically equivalent (or demographically exchangeable) and thus provide another example of a kind of distinction that may be phylogenetically significant, but ecologically irrelevant. On the other hand it is possible that behavioural distinctions within a phyletic taxon, perpetuated by lineages of cultural descent, might provide essential distinctions from an ecological perspective. It is at least a theoretical possibility that a group of organisms might require radically diverse classification from phyletic and ecological perspectives. Perhaps a population of rats, consisting of several related species, can be divided into scavengers, insectivores, herbivores, and so on in ways that do not map neatly onto the division between evolutionary lineages. Ecology may therefore, in principle at least, require either coarser or finer classifications than evolution, and it may need to appeal to classifications that cross-cut phyletic taxa.[9]

This distinction leads me to the third objection to pluralism, the metatheoretical desirability of a monistic taxonomy. Here, it is relevant to distinguish two possible aspects of pluralism. One might be a taxonomic pluralist because one believes that different groups of organisms require different principles of classification, or one might be a pluralist because one thinks that the same group of organisms require classification in different ways for different purposes. Monistic objections to the first kind of pluralism seem to me to have no merit. Taking the extreme case of bacterial taxonomy, there seem to be very good reasons for doubting the possibility of a phylogenetic taxonomy. The various mechanisms of genetic transfer that occur between bacteria suggest that their phylogenetic tree should be highly reticulated, and standard concepts of monophyly have little application to such a situation. The significance of bacteria as pathogens, symbionts, or vital elements of ecosystems make the goals of classification quite clear in many cases regardless of these problems with tracing phylogenies. Of course, it is possible that new insights into bacterial evolution might nevertheless make a phylogenetic taxonomy feasible. But no vast theoretical problem would be created if bacterial taxonomy appealed to different principles from those appropriate, say, to ornithology.[10] In this sense, the assumption that there is *any* unitary answer to 'the species problem' is no more than an optimistic hope. The suggestion that the use of different taxonomic principles might lead to serious confusion is absurd. It is of course possible that an ornithologist might mistakenly suppose that a bacterial species name referred to a

---

[9] Some more realistic examples have been discussed by Kitcher (1984).
[10] For references to bacterial taxonomy and brief discussion, see n. 3.

monophyletic group of organisms, just as it is possible that a nuclear physicist might suppose that the moon was a planet. Not every possible misunderstanding can be forestalled.

The danger of confusion is a more plausible concern regarding the idea that the same organisms might be subject to different principles of classification for different biological purposes. In one sense, I am happy to agree that this type of confusion should be avoided. It would be undesirable for a particular species name, say *Mus musculus*, to be variously defined and to have varying extensions according to the taxonomic theory espoused by various authors. We should aim to agree as far as possible which organisms are house mice. In the concluding section of this paper, I explain how I think such species names should be understood. If, to recall my hypothetical example about rats, it proves useful to treat scavenging rats as a basic kind in some ecological model, it would be misguided to insist that scavenging rats constitute a species. Equally clearly, however, this concession to standardized terminology does not at all require that all species names be conceived as answering to the same criterion of what it is to be a species. The other consequence of insisting on an unambiguous interpretation of particular species names is that we cannot assume a priori that the canonical taxonomy incorporating standard species names will be suitable for all biological purposes. The question here is, again, an empirical one that depends ultimately on how orderly biological nature turns out to be. If it should prove to be disorderly in the relevant sense, then biology would prove to be a more complicated discipline than is sometimes assumed. But once again I cannot see that any unavoidable confusion need be introduced.

Conclusion: A Case for Taxonomic Conservatism

Many taxonomists and almost everyone who uses the results of taxonomic work have complained about the genuine confusion caused by changes in taxonomic nomenclature. Some of these changes seem entirely gratuitous—for example, changes in the names of taxa grounded in the unearthing of obscure prior namings and in appeals to sometimes esoteric rules of priority. Other changes are more theoretically based adjustments of the extent of particular taxa. Many such theoretically motivated changes have been alluded to in this paper. BSC-committed theorists will urge that discoveries of substantial gene-flow between otherwise apparently good species should lead us to apply one species name to what were formerly considered several species. Phylogenetic taxonomists will want to amend the extensions of any higher taxa that fail their favoured tests for monophyly, and strict cladists will promote the breaking up of prior 'species' into various smaller units when their favoured criteria for lineage splitting demand it.[11] Less theoretically committed taxonomists may promote the splitting or lumping of higher taxa on the basis of general principles about the degree of diversity appropriate to a particular rank.

There is no doubt that the taxonomic system we now possess is a highly contingent product of various historical processes. Walters (1961) gives a fascinating account of how the size of angiosperm families and genera can very largely be explained in terms of earlier biological lore available to Linnaeus. Considering the data collected by Willis (1949) in support of the idea that the large families—families, that is, with large number of genera—were those of greater evolutionary age, Walters argues compellingly that the data much more persuasively support the hypothesis that larger families are those that have been recognized for longer. Very crudely, one might explain the point by arguing that the existence of a well-recognized type provides a focus to which subsequently discovered or distinguished types can be assimilated. Thus, plants of ancient symbolic significance, such as the rose and the lily, have provided the focus for some of the largest angiosperm families, Rosaceae and Liliaceae. Walters makes the suggestive observation that even Linnaeus, recognizing the similarities between the rosaceous fruit trees, apple, pear, quince, and medlar (*Malus, Pyrus, Cydonia*,[12] and *Mespilus*), attempted to unite them into one genus, *Pyrus*. This attempt was unsuccessful, however, presumably because of the economic significance of these plants, and

[11] De Queiroz and Gauthier (1990, 1994) claim that taxonomic changes they advocate will promote constancy of meaning, or definition, for taxa. Mammalia, for example, should be defined as the set of descendants of the most recent common ancestor (i.e. ancestral species) of monotremes and therians. The extension of such a term, however, will be constantly revisable in the light of changes in opinion about the details of evolutionary history. From the point of view of the consumer of taxonomy, at least, I suggest that constancy of extension is surely more valuable than constancy of definition.

[12] Subsequent to Walters's paper, the quince appears to have been reconceived as *Chaenomeles* (though not unanimously according to the few sources I consulted on this matter). This reconception effects a conjunction with the ornamental flowering quinces. One might speculate that the increasing obscurity of the quince as a fruit might have exposed it to this annexation, which one doubts could have happened to the apple.

modern practice has reverted to that of the seventeenth century. Walters comments: 'Can we doubt that, if these Rosaceous fruit trees had been unknown in Europe until the time of Linnaeus, we would happily have accommodated them in a single genus?' A general feature of Walters's argument is that our taxonomic system is massively Eurocentric. The shape of taxonomy has been substantially determined by which groups of plants were common or economically important in Europe.

The crucial question, of course, is whether this bias is a matter for concern and a reason for expecting wholesale revision of our taxonomic practices. To answer this question, we must have a view as to what taxonomy is for, and we come back to the major division introduced at the beginning of this essay: should we see taxonomy as answering to some uniform theoretical project or more simply as providing a general reference scheme to enable biologists to organize and communicate the wealth of biological information? The central argument of this paper is that the more we have learned about the complexity of biological diversity, the clearer it has become that any one theoretically motivated criterion for taxonomic distinctness will lead to taxonomic decisions very far removed from the desiderata for a general reference scheme. Of course, the contingencies of taxonomic history will no doubt have led, in many instances, to a scheme that is less than optimal even as a mere device for organizing biological information. On the other hand, in the absence of a theoretical imperative for revision, it is essential to weigh the benefits of a more logical organization of diversity against the costs of changing the extensions of familiar terms. My intuition is that on this criterion taxonomic revisions will seldom be justified.

We might begin by recalling part of Huxley's account of the function of classification (in the epigraph to this essay): to facilitate the operations of the mind in clearly conceiving and retaining in the memory the characters of the objects in question. Plainly to the extent that taxonomic names are undergoing constant modification, what any one person 'conceives and retains in the memory' will be potentially incommunicable to others, and the possibility of reliably adding further information obtained from the work of others will be constantly jeopardized. This is not to say that taxonomic revision is never justified. If a species is included in a genus in which it is highly anomalous, and if that species is much more similar to other species in some other genus, then the goals of organizing information will be better served by reassigning it. It is of course also true that monophyletic taxa will tend to be more homogeneous than polyphyletic taxa, and that in paraphyletic taxa—taxa in which some of the descendants of the common ancestors of a particular taxon are excluded—there will often be a case, on grounds of similarity, for including the excluded parts of the lineage. My point is just that these consequences rather than monophyly itself should provide the motivation for taxonomic change, and the benefits of such change must be weighed carefully against the potential costs. In this weighing process, the presumption that taxon names retain constant extension should probably be kept as strong as possible to maximize the ability of biologists to maintain reliable and communicable information.

To take perhaps the most familiar example, it seems to me that there is no case at all for revising the class Reptilia (reptiles) to include Aves (birds). This move is required by a strict cladistic concept of monophyly because it is believed that birds are descended from ancestral reptiles. We cannot exclude these avian ancestors from the class that includes modern reptiles because crocodiles, still classed as reptiles, are believed to have diverged from the main reptilian lineage earlier than birds did. The fact remains, however, that most zoologists, I suppose, would consider crocodiles much more like other reptiles than either is like any bird. The attempt to convince the learned or the vulgar world that birds are a kind of reptile strikes me as worse than pointless. It may be said that the only important claim is that Aves should be classified as a lower-level taxon included within Reptilia, and that this classification has nothing to do with our common usage of the terms *reptile* and *bird*. Although it is certainly the case that scientific taxonomic terms frequently differ considerably from apparently related vernacular terms, this differentiation is a source of potential confusion that should not be wilfully exacerbated (see Chapters 1 and 2 in this volume). It is also unclear what advantage is to be gained from insisting on such a revision. All evolutionists, I suppose, are likely to be familiar with recent thinking on the historical relationships within the main groups of vertebrates, and if they are not, their ignorance is not likely to be relieved by terminological legislation. Similarly, experts on smaller groups of organisms will presumably be familiar with current thinking on phylogenetic relationships within those groups. To celebrate every passing consensus on these matters with a change in taxonomic nomenclature is an inexcusable imposition of a particular professional perspective on the long-suffering consumers of taxonomy outside these phylogenetic debates.

In conclusion, I am inclined to dissociate myself from the strongest reading of the taxonomic pluralism I advocated earlier (1993; see also Kitcher 1984). In view of the limited success of theoretical articulations of the species category, it would seem to me best to return to a definition of the species as the basal unit in the taxonomic hierarchy, where the taxonomic hierarchy is considered as no more than the currently best (and minimally revised) general-purpose reference system for the cataloguing of biological diversity. This system should provide a lingua franca within which evolutionists, economists, morphologists, gardeners, wildflower enthusiasts, foresters, and so on can reliably communicate with one another. Where special studies, such as phylogeny, require different sets of categories, it would be best to avoid using the term *species* (the desirability of rejecting this concept is sometimes asserted by evolutionists). Of course, such specialized users will be free to advocate changes in taxonomic usage, but should do so only in extreme circumstances. Although I am inclined to doubt the desirability of a pluralism of overlapping taxonomies, a general taxonomy will evidently draw broadly and pluralistically on a variety of considerations. Perhaps the most important will be history, not an unattractive idea in a science in which evolutionary thought is so prominent: a goal of general taxonomy should be to preserve the biological knowledge accumulated in libraries and human brains as far as possible. In addition, there would be a range of the morphological, phylogenetic, and ecological considerations that have figured in various monistic attempts to define the species. The importance of these considerations may vary greatly from one class of organisms to another. My feeble monism is my recognition of the importance of such a general reference system. My recognition of the likelihood that different enquiries may need to provide their own specialized classifications and my tolerance of diverse inputs into the taxonomic process will leave serious monists in no doubt as to which side I am on.

The position I am advocating provides, incidentally, a quick and possibly amicable resolution to the species as individuals debate. Species, I propose, are units of classification and therefore certainly not individuals. Lineages, on the other hand, are very plausibly best seen as individuals. Often, it may be the case that the members of a species (or higher taxon) are identical to the constituents of a lineage, but of course this coincidence does not make the species a lineage. And it is doubtful whether all species, or certainly all higher taxa, are so commensurable with lineages.

Resistance to or even outrage at the kind of position I am advocating may derive from the feeling that I am flying in the face of Darwin. Darwin, after all, wrote a well-known book about the origin of species, and he was writing about a real biological process, not a naming convention. Of course, the problem is that Linnaeus (or for that matter Aristotle) also talked about species and had in mind kinds, not things. Arguably, the tension between these two usages is at the root of the great philosophical perplexity that the concept of species has generated in this century. In arguing for reversion to the earlier usage of the term *species*, I am at least honouring conventions of priority. What I am proposing, however, is not much like a Linnaean taxonomy either. As many have observed, Darwin forced us to give up any traditionally essentialist interpretation of taxonomic categories and even any objectively determinate taxonomy. But almost a century and a half of biological work in the Darwinian paradigm have also shown us that evolution does not reliably produce units of biological organization well suited to serve the classificatory purposes for which the concept of species was originally introduced, so perhaps rather than a reversion to Linnaeus, it would be better to see my proposal as a quasi-Hegelian synthesis. At any rate, if I seem to have been implying that Darwin may have been responsible for introducing some confusion into biology, I am sure no one will take this as more than a peccadillo in relation to his unquestionably positive contributions.

4

In Defence of Classification

Introduction

A topic surprisingly little discussed in recent controversies about the nature of species is classification. By that I mean the practical activity of assigning the vast numbers of organisms in the world to particular kinds. One reason for this neglect, perhaps, is that writers on species are often theorists rather than the practitioners who actually classify organisms; and perhaps in some cases these two personae coexist in the same human body without communicating a great deal with one another. Or perhaps it is because many writers on this topic have reached the conclusion that species aren't kinds at all, but individuals, so that the nature of species can have little to do with the activity of classification. At any rate, it has become common to treat species primarily as units of evolution rather than units of classification: an influential recent anthology on the topic bears the title *The Units of Evolution* and the subtitle *Essays on the Nature of Species* (Ereshefsky 1992b).

It is of course possible that the units of evolution could also serve as units of classification, and that this is in fact the case seems to be very widely assumed by writers on the topic. But, and this is the first thesis I shall defend in this paper, units of evolution are far too diverse a set of entities to provide a useable partition of organisms into classificatory units. If this is correct we need to decide whether the term species should be applied to units of evolution or units of classification. And here, if anywhere in biology, it seems to me we should honour conventions of priority. Species were understood to be basal units of a classificatory system millennia before anyone had thought to formulate a theory of evolution. And to this day almost everyone who is not professionally involved in theoretical systematics assumes that the species category is a classificatory concept. So I con- clude that species should be treated as units of classification not units of evolution.

As arguments for treating species as units of evolution are generally concerned to stress, evolution is the theory that dominates contemporary biology. As the frequently quoted title of a paper by Theodosius Dobzhansky (1973) puts it, 'Nothing in biology makes sense except in the light of evolution'.[1] Sometimes this seems to be taken as sufficient grounds for concluding that species must be the units of evolution. We do, after all, generally classify the objects with which a scientific theory deals in relation to the demands of that theory. But this argument, even when more thoroughly fleshed out, is unconvincing. Classification in biology has a life of its own. Biologists in areas only tangentially connected to evolutionary theory, such as ecologists, ethnobotanists, or ethologists, need to classify organisms, as do foresters, conservationists, gamekeepers, and herbalists. As will be discussed below, for many, perhaps even most groups of organisms, evolutionary considerations are of little or no use for classificatory purposes. And finally birdwatchers, wildflower enthusiasts, or just biologically engaged members of the public, may choose to classify organisms, even if they do not need to do so. These diverse groups of people require workable classifications that enable them to communicate among themselves and to members of other such groups, record information about natural history, and so on. If, as I argue, units of evolution inadequately meet these needs they must be distinguished from units of classification. Furthermore, it follows that in biology classification and theory are more autonomous from one another than seems to be the case in most parts of science. The fact that classification cannot, at least, be closely tied to the central theory of biology leaves room for the thoroughly pragmatic and pluralistic approach to biological taxonomy that I shall advocate.

I have mentioned the thesis that species are individuals. This thesis is based on compelling arguments that the units of evolution are individuals. If it were possible to equate the units of evolution with the units of classification then it might be legitimate to conclude from this, somewhat paradoxically, that the units of classification were individuals. But since this is not possible, there is no temptation to embrace this paradoxical conclusion. There are, on the other hand, various concepts more explicitly tied to the theory of evolution that refer to groups of organisms, for example 'population', 'lineage',

---

[1] Actually I suspect that much of molecular biology would make perfectly good sense if biological organisms had been deliberately designed by God or Martians. And arguably molecular biology is currently a rather more active field (though not, perhaps, a theory) than is the study of evolution.

and 'clade'. The arguments that species are individuals show un-problematically that we should treat these terms as applying to individuals. These points will be developed in more detail later in the paper.

Units of Evolution and Units of Classification

I have referred above to the question of whether units of evolution and units of classification might be identified with one another. But if, as I have acknowledged, the units of evolution are individuals, and since, as seems no more than a matter of definition, units of classification must be kinds, it is not at all clear what such an identification could mean. What is certainly possible is that one set of organisms could simultaneously constitute all and only the members of a kind, and all and only the parts of an individual. As a matter of fact I take it that this is often the case: the members of a species do often constitute a unit of evolution.[2] But putting matters this way should also make it clear that once we distinguish these two kinds of units, and recognize that they belong to quite distinct ontological categories, it would be strange to insist that such coincidences must always occur. If the sets consisting of those organisms that were part of a distinct unit of evolution in fact turned out to serve well as kinds from the point of view of biological classification, then it might be very convenient to make such coincidence a rule of classificatory practice. But, as I shall now explain, units of evolution are often thoroughly ill suited for this additional task.[3]

What are units of evolution? What kind of entities, that is to say, evolve? Individual organisms develop, but are not said to evolve. Evolution occurs when the properties characteristic of the individuals in a phylogenetic lineage change over time, so we can say that the units of evolution are lineages of some kind. The problem, then, is which lineages are units of evolution. A natural suggestion is that to be units of evolution lineages must be *coherent*. We don't consider humans and chimpanzees to be part of the same unit of evolution, because we believe that these two groups are now evolving independently

---

[2] Strictly, such identities are probably only approximate, as questions about the limits of a species and about the origins of a lineage may be answered differently.
[3] A more detailed argument to this effect is developed in Ch. 3 in this volume.

of one another. Thus primates, or mammals, do not appear to be units of evolution. At this point it is easy to see the appeal of the so-called Biological Species Concept (BSC). According to the BSC, a species is a group of organisms connected by reproductive relations and reproductively isolated from all other organisms. These conditions provide a satisfying account of the coherence of a lineage, and motivate the idea that the division of a lineage into two occurs when reproductive isolation is established between two parts of a lineage, paradigmatically between the main body of a species and a peripheral, geographically isolated, population.

The BSC dominated systematic theory in the early years of the Modern Synthesis, and provided a tight conceptual link between evolutionary theory and taxonomy. Unfortunately in the last few years its limitations have become increasingly clear. As was recognized from its early days, it provides no account of the taxonomy of asexual organisms, and hence of all organisms for the greater part of the history of terrestrial life. Since asexual organisms certainly evolve and certainly require classification, this is a far more serious obstacle to the project of unifying evolutionary theory and taxonomy than has often been acknowledged. But even restricting our attention for the moment to sexually reproducing organisms there are very serious problems with the BSC. Recent work suggests that reproductive isolation is neither necessary nor sufficient for the existence of morphologically coherent lineages. The classic illustration of this is found in American oaks (genus *Quercus*). Here it seems that distinct morphological species have coexisted for substantial periods of evolutionary time while undergoing significant genetic exchange through hybridization (Van Valen 1976).[4] It is of course possible to respond to this difficulty by saying that this merely shows that these trees constitute one morphologically diverse species. But this illustrates the problem that I shall argue recurs again and again in the marriage of theory and taxonomy: theoretical considerations mandate a clearly suboptimal taxonomy. To the ecologist, forester, or even landscape gardener, the kinds in question are

---

[4] Mayr (1996: 265) notes this difficulty, and suggests that it is sufficient to require only that there are properties of individuals that will prevent complete fusion of the populations. But this, surely, requires a criterion for distinguishing fused and non-fused populations, a criterion that will do all the work in deciding whether or not there is a real division between species. Magnus (1998) describes a hypothetical process by which speciation may occur sympatrically, suggesting that reproductive isolation is necessary neither for the continued coherence of a species, nor even for its origin.

naturally and appropriately treated as distinct species.[5] A somewhat different kind of problem arises with large groups of organisms in which there are few if any sharp discontinuities, either reproductive or morphological. An example is the genus *Rubus*, the brambles and their relatives, which is distributed over most of the earth. Here there may be no theoretical objection to treating such a group as a single enormous and morphologically diverse species. But given the morphological diversity there are surely pragmatic reasons for applying some kind of classification. The greengrocer and the *patissier*, at least, will want to distinguish the blackberry, the loganberry, and the raspberry, not to mention many other, more obscure, named fruits.

The insufficiency of reproductive isolation to account for the distinctness of species is illustrated by the fact that morphologically homogeneous species often seem to exhibit surprisingly little genetic interchange (Ehrlich and Raven 1969). In fact species often exist as numerous populations isolated from one another by impassable geographical obstacles, and yet appear to maintain their morphological integrity. This suggests that genetic interchange is only one among a range of possible causes of the coherence of a species (see Templeton 1992), once again emphasizing that reproductive isolation is unnecessary. Of course the existence of such additional mechanisms is also indicated by cases such as the oaks, in which coherence appears to continue despite genetic introgression. The case of isolated populations introduces another important consideration. While there is no doubt that geographical isolation in some cases leads to the inception of a new distinct lineage, it may also be followed by reintegration with the main body of the species. Even if the isolated population has for a time undergone distinct evolution, the barriers to genetic exchange may be removed before the evolution of any feature that would provide a barrier to sympatric reproduction.[6] Hence what is at one time a coherent and distinct unit of evolution may merge with

[5] Stebbins (1987: 198) notes that even more traditionally construed species often provide too coarse a classification for foresters, as these may include trees with quite different ecological properties and requirements. It is interesting to compare Sterelny's (1999) suggestion that the term 'species' be reserved for metapopulations of populations occupying ecologically diverse situations.

[6] Another way of looking at this case would be to suggest that reproductive isolation might indeed be a sufficient condition to constitute a unit of evolution. However, it would be highly undesirable to be committed to calling every temporarily isolated part of a species a distinct species, thus showing again that species should not be identified with units of evolution.

another unit of evolution. Units of evolution do not form a treelike structure, but are at least partially reticulated. Since evolutionarily grounded taxonomies are invariably presented in an unreticulated treelike structure, this means that the question of whether a unit of evolution counts as coherent depends on the question of whether it will continue to evolve independently, or eventually remerge with the parent species. It is thus a question that in many cases will be operationally undecidable. It hardly needs emphasizing that this is a disadvantage for a practically applicable taxonomy. A final point is that having recognized the likelihood that there are mechanisms other than reproductive isolation that can produce evolutionary coherence, it is entirely possible that there may be subspecific units of evolution deriving coherence from such mechanisms even when these are sympatric with the remainder of the species.

Two general problems emerge. First, if we attempt to apply a rigid criterion such as reproductive isolation we will end up with species which range in size from enormous groups of (currently recognized) species such as the brambles or the oaks, to groups consisting of a handful of organisms. Second, we begin to see that the very idea of a unit of evolution is much vaguer than might first have been supposed. The concept of 'reproductive isolation' suggests the picture of evolutionary change flowing down sharply defined channels, branching at well-defined nodes—and naturally identifies units of evolution as lengths of channel terminating at nodes. A more realistic picture would be a river estuary at low tide. We find large streams of water and many side streams, some petering out, others rejoining a main channel or crossing into a different channel, and a few maintaining their integrity to the ocean; there are islands around which streams flow and then rejoin; eddies and vortices; and so on. Some parts of the general flow are naturally and coherently distinguishable, and it is easy enough to recognize parts of the pattern that are definitely not parts of the same 'unit of flow'. But in between, there are many cases where any such distinction into discrete units would be largely arbitrary. To quote a prominent defender of an evolutionarily grounded view of the species:

most species are more like slime molds and sponges than like highly organized and tightly integrated multicellular organisms—at least in terms of their individuality. Not only can almost any part of a species give rise to a new lineage, but those new lineages also commonly reunite after separating. Consequently there will be many cases in which it will be difficult to determine the precise number and boundaries of species . . . (de Queiroz 1999: 79)

The moral de Queiroz draws from this point is that it should no more lead us to doubt the reality of species than similar problems about the individuation of organisms should lead us to doubt the reality of organisms. I agree, provided that what is under discussion is the existence of lineages, or units of evolution. Whether the situation described is likely to provide us with a suitable basis for a classificatory system is quite another matter.

I introduced the present discussion by noting that a standard approach to these questions is to take it as a matter of definition that species are what evolve, and thus to take the answer to the question *what lineages are the units of evolution* as also providing the answer to the question *what groups of organisms are species*. What I hope to have shown is that this approach is probably unworkable, and to the extent that it *is* workable, will lead to a taxonomy that is seriously at odds with the desiderata of usefully sorting organisms into kinds in ways that will serve to convey and store information about the biological world.

One final worry about the divorce of species from units of evolution is not so much an argument as a matter of historical veneration. Darwin, the most venerable figure in the history of biology, did, after all, write a rather well-known book on the origin of species, and convinced most of those worth convincing that the origin of species was that they had evolved. So is the denial that species are what evolve not flying in the face of Darwin? In response to this I would say, first, that I am confident that my argument here contradicts nothing of the substance of Darwin's contribution. The greatest part of that contribution, I suppose, was the provision of a naturalistic explanation of adaptation, something almost wholly irrelevant to the present topic. He also had much to say about the origin of diversity, much closer to the present topic. But as far as I can see my main disagreement here is semantic: I don't disagree about the kind of things that evolved, only whether these should be referred to as species. Although, as I discuss in the penultimate section of this paper, I do not take semantic debates to be trivial, I do suggest that this semantic dissent falls well short of sacrilege. In the conclusion of the paper I shall, as is customary in theoretical discussions of biology, suggest that the position I take is ultimately more faithful to the spirit of Darwin than are the positions I oppose.

I have not tried to discuss in detail the many specific proposals for evolutionarily grounded species concepts that have been developed as successors to the BSC, though I have discussed some of the most prominent elsewhere (Chapter 3 in this volume). Here I mean only to give a general sense of why I think such projects are basically misguided. I now turn to a consideration of an alternative response to this situation.

Pluralism about Species and the Aims of Classification

I have no wish to deny that concordance with phylogeny is one desideratum in the provision of a taxonomy. In many of the phyla of greatest immediate interest to us, mammals or birds, for instance, it is probably reasonable to expect that taxonomy should recapitulate phylogeny. These are the taxa in which species are, in the main, more 'like highly organized and tightly integrated multicellular organisms'. Where such recapitulation is possible we should expect that evolutionary species should coincide with morphological distinctness. Michael Ruse has very optimistically claimed that 'the different ways of breaking organisms into groups . . . coincide! The genetic species is the morphological species is the reproductively isolated species is the group with common ancestors' (Ruse 1987: 237). No doubt this is often the case with these relatively well-behaved taxa. But it is surely not the case for most of the biological world. Hence I take this concordance to be a desideratum, but only one desideratum among several, and by no means the universally decisive one.

Once it is recognized that there is no theoretical grounding for a classificatory system that will universally or even generally provide a practically applicable taxonomy, we are free to embrace taxonomic pluralism. Approaches to classification will vary from one group of organisms to another, and we should allow experts in particular phyla to decide the most appropriate way of classifying a particular domain. It is worth while, however, to insist on the importance of certain desiderata derived from considerations of the nature and function of classification itself, an insistence that only becomes possible once we are freed from the prejudice that the ultimate criteria of classificatory excellence must be imported from some external source, evolutionary theory. In many parts of biology, for example bacteria and many orders of flowering plants, it is doubtful whether any evolutionarily grounded taxonomic scheme will be feasible, and it may be necessary to resort to morphology. This does not, of course, mean that we must adopt the discredited pheneticist notion of absolute similarity. Morphological features must be selected for taxonomic purposes, perhaps on the basis of biological salience (as Linnaeus considered the sexual parts of flowering parts to have particular biological importance) or perhaps just on the basis of suitability for taxonomic purposes themselves. This leads us again to the question of what the intrinsic virtues of a taxonomic scheme should be. Some obvious ones can readily be identified:

Comprehensiveness. A requirement of an adequate system of classification is that everything within its domain—in the present case every organism—should be included within it. This does not of course preclude the possibility that some allocations of organisms to taxa will be more or less arbitrary; but in that case there will at least be an approximate boundary between kinds that determines some possible decisions about the kind to which any organism belongs. There might perhaps be no fact of the matter whether I am looking at an instance of *Rudbeckia hirta* or of *Rudbeckia triloba*; or whether, perhaps, it is some kind of hybrid between the two. But it certainly isn't a fish, or even an oak tree. And if I tell you it's a black-eyed Susan you shouldn't be seriously misled. The condition that something useful can be said about any organism seems a minimal requirement for an acceptable taxonomy.[7]

Suitably sized base-level groups. One of the main defects with reproductively or phylogenetically determined groups is that these tend to generate groups differing in size by many orders of magnitude. This defect can perhaps be remedied by the kind of moderate pluralism proposed by Mishler and Donoghue (1982), who insist that species should be monophyletic, while adopting a pluralistic attitude to the question of what monophyletic units should count as species. This remedy may make one wonder, however, what virtue is being served by the continued insistence on monophyly. The criterion of suitable size is necessarily a vague one. Bitterns are rarer than blackbirds, but this should not motivate the attempt to divide blackbirds up into lots of geographic races about as numerous as are bitterns. Despite this vagueness, however, it is clear enough that some such desideratum

---

[7] Sterelny (1999), however, argues that only some kinds of organisms are organized into species. I take this as illustrating my claim that species thus conceived are inadequate as a basis for taxonomy. Sterelny's position is discussed briefly below.

is violated by the suggestion that there is only one species of deciduous oak in America, or that common types of bacteria should be divided into innumerable species the vast majority of which go extinct long before anyone bothers to distinguish them.

Reasonably homogeneous groups and *Reasonably sharp boundaries between groups*. I treat these together because the limits to their attainability can be said jointly to define the problem for biological taxonomy, and the respect in which it differs from traditional essentialist taxonomy. An ideal taxonomy would distinguish entities as members of wholly homogeneous classes, which are reliably distinguishable from the equally homogeneous members of every other class. This ideal would have been achieved by a taxonomy based on the essential property or properties of members of the taxon, if only there had been such properties. We know that the phenomena of biological variation, hybridization, and more or less imperfect separation between clades make such a taxonomy unachievable in biology. But the best must not be allowed to be the enemy of the good. These characteristics continue to remain ideals of classification, simply because in obvious ways they make taxonomies easier to use. The ideal of homogeneity is another part of the reason why groups such as the oaks and blackberries do not provide good base-level taxa. Unlike homogeneity sharp boundaries are always attainable. In many familiar cases they simply exist between fully isolated and phylogenetically distinct groups and can be identified through several or many different possible criteria. But where this is not the case they can still be defined. One can, for example, distinguish blackberries as belonging to a particular species by drawing geographic lines and defining anything to the south of the line as belonging to one species, anything to the north to another. Whether such a consequence as different parts of the same bush belonging to different species would be a price worth paying for this advantage is another matter.

Taxonomic conservatism. A quite different kind of desideratum for a taxonomic scheme is that it should be as stable as possible. Consumers of taxonomy, people who use the results of expert taxonomists in their everyday interactions with the biological environment, often complain about the constant tendency of biologists to change classifications, thus making existing taxonomic expertise ineffective. The main relevance of this to the present topic is the argument it provides, from the perspective of a pragmatically useful taxonomy, against the assumption that taxonomy should be theory-driven. The nature of species is a notoriously contentious matter, so it is hardly surprising that biologists driven by one conception of the species should feel compelled to revise the taxonomies of their predecessors, often from a very different theoretical perspective. More important still, commitment to a uniform theoretical view of the species category implies that new biological information will constantly motivate changes to taxonomy. If, for example, one is committed to the BSC, then the discovery that particular populations do or do not interbreed will force revisions of the application of species names. The advantages of taxonomic stability are obvious and considerable, maximizing the stability of detailed biological knowledge whether stored in libraries or human brains. The fact that a pragmatic and pluralistic conception of taxonomy is able to set a much higher standard for taxonomic revision is a considerable advantage for such an approach.

I have not, of course, said anything about the biological rationale for particular conceptions of taxonomy. To the extent that a biological rationale is seen as forcing taxonomy into the theoretically sanctioned form of monism, any such rationale is opposed to the position I am defending. However, it is quite consistent with my view that theoretical biological considerations should be locally important in determining how particular groups of organisms should be classified. As noted above, it is a fine thing if a taxonomy can coincide with a plausible phylogeny. For groups such as birds and mammals this is a realistic goal, and may even be a reasonable constraint on taxonomic practice. However, it is almost certainly not a generally feasible constraint. It almost certainly cannot be applied to prokaryotes, hence to the first half of evolutionary history, and perhaps is only applicable to a few extremely highly evolved and sophisticated groups of metazoans. The only possible unifying conception of the species is as a unit not of evolution but of classification.

This is the view of species that I propose. Species are a unified category only in the very weak sense that they are (by definition, I suggest) the basal units of classification. But the criteria used for distinguishing species from one another, and the theoretical rationale, if any, for such distinctions, should be allowed to vary promiscuously from one taxon to another.

Units of Evolution as Individuals

One of the most widely discussed theses in the recent philosophy of biology is the idea that species are individuals.[8] Provided it is assumed that species are units of evolution, this thesis is extremely compelling. To begin with the most obvious and, I believe, the fundamental point, evolution is a process that something, presumably, undergoes. But it is doubtful whether it even makes sense to speak of a *kind* undergoing such a process. A kind can be the object of mental processes; it can be contemplated or its members can be counted. But it isn't the kind of thing that can undergo processes in the world outside. It is true that collectives consisting of numbers of individuals can undergo processes. A crowd can become unruly, or a herd of cows stampede. But of course a crowd or a herd is not a kind. One is a member of a crowd by virtue of the relations (mainly spatial) that one bears to other members, not by virtue of any intrinsic properties one may have. The possible, if unlikely, crowds in *Star Wars* movies consist of exceedingly diverse entities; and the identical twin of a member of a crowd is not, by virtue of similarity, a member of the same crowd. Moreover, from the point of view of the theory of crowds, crowds are individuals, and members of crowds are parts of those individuals. It is extremely plausible that the same should be said for species *qua* units of evolution. If, somehow, the kind Horse were evolving, then possible horses in distant galaxies would be participating in the same process as would merely terrestrial horses. For presumably terrestrial and extra-galactic horses bear the same relation to the kind Horse. But of course this is nonsense, and the evolution of horses on earth has no bearing on whatever horses there may be elsewhere in the universe. An obvious explanation is that horses are parts of the evolving unit rather than members of an evolving kind. And it is individuals, never kinds, that have parts.

Another commonly cited argument is that kinds, or at any rate natural kinds—which species would presumably be if they were kinds—are the subjects of laws of nature; and there are no laws, it is said, about species. I am rather less sure about this argument, if only because it seems to me so far from clear what a law of nature is. Many laws of nature appear to be probabilistic, and the fact that there is

---

[8] The most detailed elaboration of this idea is by Michael Ghiselin (1997), who first introduced the thesis in the 1960s.

always or almost always variation among the members of a species does not obviously preclude there being probabilistic laws about such members. Matters are murkier if we take it for granted that the question is whether there could be laws applying to the parts of a unit of evolution. It is often supposed that this must be precluded by a restriction on laws that they make no reference to any spatio-temporally specific individual. But the motivation for this restriction is obscure. Kepler's laws, which refer to a mere handful of celestial objects, were once the best laws of nature going. They are not, of course, exactly true, but I suppose they might have been, and anyhow the laws that impress us most are not exactly, or even roughly, true most of the time (Cartwright 1983). There is, on the other hand, an obvious reason from my perspective on the matter why we should not expect very reliable or fundamental laws about the members of species. This is simply that species are distinguished primarily for the purposes of classification, not generally for deeper theoretical reasons. If species are divorced to some degree from biological theory, then they will perhaps be equally distanced from laws. But then it is the vast diversity of biological phenomena that forces us to accept a classificatory system divorced from theory. So it may be that species provide the subject matter for laws as exact and fundamental as are available for biology at the organismic level.

My main point about these arguments, however, is that they start from the assumption that species are what evolve, that they are the units of evolution. And this is what I want to deny. I do not deny, on the other hand, that the things that evolve are historically specific individuals. If these are not species, what are they? The most plausible answer, and ultimately the one I take to be correct, is that the units of evolution are *lineages*. Lineages in general are sequences of entities related by ancestry and descent. The relevant entities in this case are populations, or groups of interconnected populations. Populations are themselves, I take it, individuals, and they are the temporal parts of other individuals, lineages.[9] The general motivation for this suggestion is that evolutionary change is thought of as accumulating within lineages as it is passed from ancestors to descendants. Lineages should be distinguished from the technical concept of a clade since clades, but not lineages, are generally defined as including all the

---

[9] A useful discussion of the concept of a lineage is provided by de Queiroz (1999), though in the context of an argument I reject, that species are segments of lineages.

descendants of a specified ancestral population. A clade, therefore, may contain distinct branches that are no longer a part of the same unit of evolution. There is no reason to think of a lineage, on the other hand, as ceasing to exist because a side-lineage achieves the status of an independent and distinct lineage.

It would be nice if the biological world divided up into fully distinct and internally fully coherent lineages. But of course it is just the failure of this occurrence that provided a fundamental obstacle to the attempt to assimilate units of classification to units of evolution. Typically we should expect to find ever smaller units of evolution —demes, temporarily isolated populations, etc.—within units of evolution, and in groups of organisms for which horizontal gene introgression is common, attempts to isolate the largest units of evolution will be to some degree arbitrary. This may be regrettable, but since the question *what are the units of evolution* is intended as properly ontological, in answering it we must attempt to record the actual state of things as accurately as possible. For the purposes of specific evolutionary investigations a degree of pluralism may well be appropriate. For some theoretical purposes it will be most useful to treat large, and perhaps only partially coherent, lineages as the units of evolution. For other purposes one might need to consider the properties of numerous and short-lived units. At any rate, whether we are considering major branches of the phylogenetic tree, local populations of particular organisms, or clones of bacteria maintaining their integrity for minutes or seconds, in so far as these are conceived as units of evolution, they are appropriately thought of as individuals.

Some Consequences

In a sense what I have been arguing is primarily a semantic point, though none the worse for that, I take it. However, a little more should perhaps be said about the rationale for the semantic decision here being advocated with regard to the word 'species'. Kim Sterelny (1999: 123) writes: 'we have to choose between the ideas that all organisms are members of some species and that being a species is an important biological property, for just as the *organism* as a grade of biological organization had to be invented, so did the *species*'. The first part of Sterelny's remark perfectly summarizes my conclusions about the impossibility of coincidence between the units of evolution and the units of classification. But whereas I take the first option, on the grounds noted above that a system of classification must be comprehensive, Sterelny opts for the second. Sterelny argues that the biologically interesting level of organization that corresponds, sometimes, to the traditional species, is a metapopulation consisting of an ecologically variable mosaic of populations. The choice Sterelny presents arises because, unsurprisingly, many kinds of organisms are not found in such biologically interesting metapopulations. But this makes Sterelny's semantic decision a bizarre one. On the one hand he offers us a term, 'metapopulation', that adequately captures the allegedly biologically interesting level of organization. If this is not specific enough, one could be invented—'ecometapopulation', perhaps. On the other hand we have a term that has been used for several millennia to refer to the basic unit of biological classification. Recognizing that in some cases the group of organisms distinguished by a traditional classificatory term coincides with a biologically interesting evolutionary individual strikes me as a very bad reason for co-opting the term 'species' to refer solely to such individuals. So I enlist Sterelny as a hostile witness in support of my claim that we must give up the idea that being a species is an important biological property.

This raises a more general philosophical issue. It is often assumed, and I have made just this assumption myself, that classification in science must go hand in hand with theory. Progress in chemistry, for instance, is immediately reflected in improved classification of material stuffs, and modern chemistry is enshrined in the periodic table of the elements, both a scheme of classification and a vehicle for a good deal of chemical theory. Pure chemicals are the best surviving candidates for the traditional conception of natural kinds demarcated by the common possession of real essences. And something similar may be seen at the other end of the spectrum of organizational complexity. If Marx's social theories, for instance, have merit, then it should be theoretically illuminating—or anyhow should once have been— to classify people as proletarians and bourgeois. But of course no one thinks that all proletarians share a common real essence. What they share is what the theory says they share, a relation to the means of production. If the theory is a good one it will be useful for certain purposes to classify people by their relation to the means of production. But for other purposes it will be more relevant to speak of their nationality, their gender, the precise industry in which they work, or many other things. Perhaps chemistry is a simple enough science that one system of classification can be suited to all theoretical enquiries, but this is patently not the case for people. In fact for the human sciences there is no general taxonomy. The question *what kind does that human belong to* is meaningless without further context. There are, perhaps, people who think there is a clear answer to such questions, namely racists. But their views have thankfully been discredited, if by no means abandoned.

A view such as Sterelny's treats biology as in certain respects like the human sciences. It assumes that there is some theoretical project towards which the classification of organisms into species is directed, and that this project is the formulation of the best possible account of evolution. Since only certain species-like groups of organisms turn out to play a coherent role in the evolutionary process, only certain organisms belong to species. Similarly if there were wild people living in the woods off berries and roots a Marxist might say that they bear no relation to the means of production and thus belong to no social class. But the analogy between the biological and the human sciences breaks down here: in biology there *is* an important role for a general classification. As I have noted above, many different concerns, both theoretical and practical, generate an interest in the classification of organisms. But unlike the situation generally supposed to obtain in chemistry, there is no uniform theoretical perspective that will generate such a classification.

It appears, then, that the relation between theory and taxonomy is diverse. In the physical sciences the two are tightly yoked together by virtue of the existence of something close to traditional natural kinds; in the human sciences, paradoxically, they are also yoked together, this time by the fact that there are many different classifications relative to many different theoretical perspectives, and that there is no need for a general taxonomy. But in biology there are no such traditional natural kinds, many theoretical perspectives, but a massive diversity of highly diverse kinds. This situation forces classification to lead a life of its own, sometimes tightly connected to important theoretical perspectives, sometimes not.

One main reason why it is thought desirable that classification and theory should be closely connected is that we hope to discover laws about the kinds of things distinguished by our taxonomies. Indeed where the relation is most intimate, as in both the examples from chemistry and the examples from the human sciences, taxonomies are developed precisely in relation to laws that are discovered, or at least proposed, about members of kinds. Indeed, the possibility of formulating such laws might well have been added to my earlier list of desiderata for a classification. This brings up again the issue of whether there are laws about the members of biological species. This is certainly too large an issue for me to address in any detail here; for one thing, the question of what constitutes a scientific law, or a law of nature, is a thoroughly vexed one. What can safely be said is that the partial divorce of taxonomy from theory in biology provides one clear reason why we cannot expect laws of any fundamental importance to govern the behaviour of members of species. Perhaps the sheer number of different species is another such reason; laws referring to any one, unless perhaps our own, will be of only the most marginal importance. On the other hand, given that homogeneity is an important desideratum of the taxa generated by a classification, we should certainly expect there to be modest but fairly reliable generalizations about the members of species. And indeed the unduly maligned domain of natural history is largely the domain of such generalizations. I have no axe to grind about whether some or all of these should be accorded the honorific title of laws. Reasons for scepticism about such laws are often grounded in the assumption that species are units of evolution, and therefore subject to diachronic change and synchronic variability. If bacteria, for example, were to be classified instead in terms of morphology or ecological role, these arguments might be less compelling. But it would still be the case that the fact that we are concerned with only a minute fraction of the organic world will render the generalizations about a particular species of extremely local significance, and seldom if ever of deep theoretical import.

Let me finally say something about the question of realism. In earlier work (Dupré 1993) I have insisted that despite the pluralism of biological classification, there is no reason to doubt that the kinds thus promiscuously distinguished should nevertheless have some reality, and indeed qualify, if in a sense rather weaker than the traditional one, for the status of natural kinds. I continue to maintain that there is no reason why the account of species currently offered should preclude their being modestly natural kinds.[10] However,

---

[10] One account of what might constitute them as natural kinds is that of Boyd (1999) and Wilson (1999).

the recognition of classification as the predominant driving force behind the distinction of species leads me to moderate my earlier promiscuous realism about biological kinds with a more promiscuous metaphysics. Often there are ways of classifying organisms in ways that correspond to modestly natural kinds, but often there may not be. Even in the latter case—bacteria, brambles, perhaps—classification must go on. So I am inclined to say that some species are real natural kinds, but many are not.

Conclusion

The assumption that species must be units of evolution is undoubtedly related to the extent to which the figure of Darwin and his decisive contribution to biology dominates the discipline. As I quoted earlier, 'Nothing in biology makes sense except in the light of evolution' (Dobzhansky 1973). I conclude with two comments about this remark. First, I am inclined to argue that the position I am defending reflects a further assimilation of Darwin's insights. It is often noted that Darwin put an end once and for all to the Aristotelian tradition of seeing biological species as natural kinds determined by real essences (Hull 1965). This naturally led to a profound concern with the question: What, then, are species? Naturally, too, biologists and philosophers have tried to find an answer in the terms of the Darwinian paradigm. Darwin offered an account of the origin of biological diversity, and it has naturally been assumed that that was an account of the origin of species, where species would turn out to be something like what the Aristotelian perspective had assumed. Such an assumption was perhaps an essential stage in the evolution of our conception of biological diversity. But post-Darwinian biology has increasingly revealed that in many parts of the biological world there really are nothing like traditional species at all. Evolution has generated highly diverse patterns of diversity, some of which involve divisions similar to, or even coextensive with, what have previously been considered species, but some of which do not. Perhaps Darwin's work should have been entitled *The Non-Existence of Species*, except that no one would have had any reason to take such a work seriously. But even if there are no species in the Aristotelian sense, and no species in the sense of real units in nature reliably produced by the evolutionary process, we still have good reasons to impose some taxonomic order on the biological world. And in view of the picture of the biological world that has developed from Darwin's theory, this process can only be achieved with a methodology that is pragmatic, pluralistic, and sometimes frankly nominalist.

My second point is related to the first, but is slightly less respectful of the Darwinian hegemony. Though I have no doubt that every domain of biology is populated by organisms that evolved, there are parts of biology upon which evolution sheds less light than is often assumed. One of these, far removed from the present topic, is human behaviour. No doubt we evolved, and in a broad sense so did our behaviour. But the thesis that this fact is the key to understanding human behaviour has generally, and is currently, generating conclusions that range from the banal to the ridiculous. Parts of taxonomy may well also turn out to be areas where Darwin's insights shed less light than is often expected. David Nanney, an expert on the protozoans, a fascinating and bizarre group of organisms noted for their tendency to provide exceptions to widely assumed biological generalizations, has lucidly described the extraordinary difficulties in providing a taxonomy of an important group of these organisms, the Tetrahymena, in the light of evolutionary principles. He remarks:

the earlier refusal of protozoologists to designate cryptic species by Latin binomials led some evolutionists to suppose that protozoologists do not understand evolutionary biology. Protozoologists, on the other hand, suspect that evolutionary geneticists do not always understand that taxonomy serves clients other than evolutionists. Different terms may be needed in different contexts. (Nanney 1999: 99)

It is part of the moral of this paper that we show no disrespect to Darwin when we recognize that sometimes there is more to biology than just the fact that organisms evolved.

III
Kinds of Kinds

5

Is 'Natural Kind' a
Natural Kind Term?

The traditional home for the concept of a natural kind in biology is of course taxonomy, the sorting of organisms into a nested hierarchy of kinds. Many taxonomists and most philosophers of biology now deny that it is possible to sort organisms into natural kinds. Many do not think that biological taxonomy sorts them into kinds at all, but rather identifies them as parts of historical individuals. But at any rate if the species, genera, and so on of biological taxonomy are kinds at all, there are various respects in which they fall short of the traditional requirements of naturalness. The members of biological taxa lack essential properties that make them members of a particular kind: any properties specific enough to belong only to members of the kind cannot be assumed to belong to all members of the kind. And if there are laws applying to members of biological taxa, they are laws of very minor and local importance and, in view of the preceding point, at best probabilistic.

My main focus in this paper, however, will not be on the nature of biological taxa. The acceptance that biological taxa are not natural kinds has not generally been taken to show that there are no natural kinds in biology. It is just that, on this view, we must seek them at a more theoretical level than is the concern of applied taxonomy. So although '*Ranunculus repens*' or '*Felis leo*' may not name a natural kind, perhaps 'species' does. Perhaps the members of *Felis leo* also belong to a kind of ecology, predators. It has been supposed that the whole lion might instantiate a fundamental natural kind of evolutionary theory, interactor; while tiny little bits of it, lurking at the centre of each of its cells, might be members of the kind replicator. And surely these little bits at least belong to kinds of organic chemistry. Perhaps grosser parts of the lion may instantiate kinds such as liver, heart, or teeth. And so on.

Why might it matter to identify such kinds? The central reason is that the prevalence of kinds is commensurate with the prevalence of laws. If there are high-level kinds of evolutionary theory, ecology, or physiology, then there will be laws about the members of such kinds. This, I take it, is the most fundamental importance of identifying natural kinds in science: they tell us what the laws of that science should govern. If, on the other hand, there are no such higher-level kinds, we should not expect there to be any wide-ranging laws either. There is no dispute, of course, that biology is full of low-level generalizations: wolves hunt in packs, or humans have an average body temperature of 36.9 degrees Celsius, for instance. The limitation of such generalizations of natural history is just that they give us no general guidance as to what to expect in different cases. If we want to move on to the hunting behaviour of crocodiles, or the body temperature of hummingbirds, we must start again from scratch. Contrast Dollo's law, which states that evolution never brings a population back to a state that it has abandoned in the past. This is a very weak law, in that it says only one thing that doesn't happen, nothing about what evolution will in fact do. And it may well prove not to be true; or if it is true, it is very probably no more than a special case of the fact that given any pre-selected feature of a population, the chance that an evolving population will reach that state is approximately zero. However, it is clear that its scope is wide: if it is true, it applies to any evolving population of anything.

In short, then, whether there are natural kinds or not in biology tells us something fundamental about the structure of biology. Biologists and philosophers of biology have been engaged for several decades in discarding the baggage of a model of science derived from reflection on physics and chemistry. This model puts at the centre of attention the search for wide-ranging laws and the natural kinds over which these laws range. The insistence that there are no natural kinds of this sort in biology is a contribution to this ongoing emancipation of biology from an inappropriate model. Indirectly, therefore, such an argument also offers support to the alternative conception of the structure of biology that has been developing over these decades, the idea that biological theory provides us not with universal laws, but rather with arrays of models that biologists learn to apply to suitable systems encountered in nature or the laboratory. My central aim in considering biological kinds in the present paper is to contribute to this major shift in the conception of biological science, and perhaps even science in general.

In the course of this paper I shall look at a few of the more plausible candidates for high-level natural kinds in various areas of biology.

As well as trying to show why these are not appropriately thought of as natural kinds in the traditional sense, I shall offer some suggestions for what might be better ways of thinking of some of these concepts. This, I hope, will provide some further insight into the characteristics of post-nomological biology.

Before doing this, however, I want to suggest a somewhat different way of viewing natural kinds. In the past I have suggested that we should favour a much weaker conception of natural kinds than has been customary. In defending a pluralism of taxonomies against the accusation of inviting a nihilistic relativism, I have insisted that relative to a sufficiently well-articulated set of aims of enquiry there may very well be, and often is, a best way of classifying the phenomena within a domain.[1] Relative to such a specific enquiry, it seems sensible to say that such a best classification distinguishes natural kinds. The pluralism, of course, implies that there will usually be no unique set of natural kinds, and different natural classifications may often overlap and cross-cut one another. The natural kinds of ecology, say, may not coincide with the natural kinds of phylogeny. But natural kinds in this sense will necessarily fail to meet traditional desiderata. They will not point towards the essence of a thing, and they will only tell how a thing should be classified for a certain purpose, not how it should be classified, period.

This view of natural kinds indicates how I propose to answer the question that forms the title of this paper. If 'natural kind' is understood in a traditional, metaphysical, sense, then the answer is presumably that natural kinds do indeed form a natural kind—the kind of kinds whose members share a common essence—but a sparsely occupied one with few or no biological kinds as members. On my preferred view natural kinds surely do not form a metaphysical natural kind; at best they might constitute an epistemological or methodological kind. But even this, I think, may tend to exaggerate the homogeneity of the kinds that fall under this higher-level kind. Anticipating a proposal I want to discuss at the lower level of biology, I think it may be better to say just that 'natural kind' is a (more or less) useful methodological concept.

---

[1] The explanation of there being such a best classification may often be along the lines of Richard Boyd's account of homeostatic cluster kinds. Boyd (1999) applies this account to biological taxa. I think Boyd's account is probably more restrictive than the view of natural kinds I advocate here, however.

Traditional Natural Kinds

Traditionally natural kinds were generally assumed to satisfy all or most of the following conditions:

1. Membership of the kind was determined by possession of an essential property or properties.

2. Members of a natural kind were the appropriate subjects of scientific laws, or laws of nature.

3. By virtue of condition 2, the properties or behaviour of the members of a natural kind were to be explained by identifying the kind to which a thing belonged, and referring to the laws governing things of that kind.

4. The conformity of members of a natural kind to laws of nature was ultimately to be explained by appeal to their essential property or properties.

5. If a thing belongs to more than one natural kind, it must be the case that the kinds to which it belongs are part of a hierarchy in which lower-level members are wholly included in higher-level members. If humans, mammals, and animals are all natural kinds, and I am a member of all of these kinds, then all humans must be mammals and all mammals animals (or some other parallel set of relations must obtain). This requirement is reflected in the idea that to say what kind a thing belongs to is to say what it really is. The only way that there can be two answers to this question is if the kind of thing something really is (e.g. human) is of a kind that is included within a higher-level kind (e.g. animal).

These conditions are interrelated in obvious ways, reflecting the fact that the traditional concept is central to a general vision of science encompassing views about laws of nature and explanation, among other things. It is, therefore, difficult to give up substantial parts of this account without abandoning the entire picture on which it is based. One might perhaps try to relax or even abandon the fifth condition, though this would present obvious difficulties when the behaviour or properties characteristic of different kinds to which a thing belonged were in conflict. One might alternatively elaborate the general account in terms of a conception of essence less rigorous than has generally been assumed, for example in terms of a cluster of properties of which only some subset would be required for kind membership. This would naturally be further articulated in terms of probabilistic rather than deterministic laws, since the aspect of the essential cluster responsible for a particular form of lawlike behaviour would be absent from a certain proportion of members of the kind.

I do not, however, intend to pursue in any detail such possible weakenings or modifications of the general picture, because it seems to me that the entire conception has been largely abandoned for most of science, and specifically for biology. (I do not want to engage here in the debate over whether the general conception remains appropriate for understanding the physical sciences.) The clearest indication of this move is in the tendency to think of science as involving not universal laws but models.[2] A model of a process under investigation will be useful to the extent that the elements of the model adequately represent the causally relevant aspects of the process. It carries no guarantee that it will be applicable to further cases. It is always an empirical issue whether other distinct processes will have sufficiently similar causal determinants to be usefully represented by the same model. The ability to model successfully a wide range of distinct though related processes is generally seen as depending on access to a variety of more or less related models, a toolkit of models, one of which will often be found capable of application to the problem in view.

This general perspective does still require supplementation by some account of the generality of scientific theory. Why, for instance, do we think of models involving sexual selection, assortative mating, heterozygote superiority, genetic linkage, and so on as all falling under the broad umbrella of the theory of evolution? What, one might say, qualifies a particular tool to belong in a particular toolkit? This seems to me an important question, but all I want to say about it here is that there is no likelihood that the answer will have anything to do with traditional natural kinds or their hypothetical essences. Following the advice of Ernst Mayr (1997), we would do much better to look at the role of fundamental concepts in uniting the various elements of the evolutionary toolkit than at natural kinds. For example, most of these models depend in important ways on the concept of fitness, or of the bases for dispositions to leave greater or smaller

---

[2] There is a large and rapidly growing literature on the role and importance of models in science, which I cannot even attempt to summarize here. This development is, of course, also related to the debate between proponents of the syntactic and semantic views of theories. This debate, however, tends to be framed in terms of technical issues that will not concern me here.

numbers of descendants. The bases of such dispositions are as varied as the biosphere, and no one supposes that there is a natural kind of fit things, or things with a fitness of 0.7, still less that there is an essential property that underlies the possession of that level of fitness.

Elliott Sober (1984*b*) has suggested that one fundamental law of evolutionary theory is the Hardy–Weinberg law, which states that for a diploid locus with two possible alleles A and a, with frequencies p and q, the distribution of the genotypes AA, Aa, and aa will be p^2, $2pq$, and q^2. The importance that Sober attributes to this law is that it is a 'zero-force law': deviations from the equilibrium relations described by the law provide evidence that some force is acting on the population. I have some reservations about this proposal, but they are not my present concern. All I want to note is that once again there are no candidates for natural kinds in the offing. Homozygote, heterozygote, gene frequency, and so on may be essential concepts in forming certain kinds of evolutionary models, but no one supposes that they name traditional natural kinds. One might perhaps see the Hardy–Weinberg law as a structural condition of all or many evolutionary models, and thus, perhaps as part of the answer to why different evolutionary models belong, as I have put it, in the same toolkit.

I have suggested that there is a weaker concept of natural kinds that may nevertheless play an important role in analysing scientific ideas, one that simply reflects the fact that from the point of view of a particular scientific enquiry there may be a right way to classify the phenomena. It will be easiest to illustrate the idea with an example. Consider a textbook example of evolutionary explanation, the prevalence of sickle-cell anaemia. This is, of course, explained as an example of heterozygote superiority. While the sickle-cell homozygote is highly deleterious or lethal, the heterozygote is less susceptible than the viable homozygote to malaria. Where malaria is prevalent, selection for the heterozygote and against the homozygote creates an equilibrium that can be expressed in terms of the Hardy–Weinberg law. From the point of view of understanding the phenomena in question we have to distinguish the three relevant genotypes. And relevant to that interest, we might say that the specific genotypes just mentioned were natural kinds. It is by virtue of the different causal properties of those kinds that the phenomenon at issue is to be explained. I do not take this suggestion to differ in any fundamental respect from the proposal above that we should think in terms of concepts rather than kinds. It is just that the application of important theoretical concepts will, among other things, serve to sort individuals into groups. The construction of models will typically involve distinguishing between the roles of members of these groups. Concepts, in short, classify things, and it is with views of classification that I am here concerned.

But this sorting of things into the kinds that address our particular causal and explanatory concerns delivers kinds that satisfy few if any of the criteria for traditional natural kinds distinguished above. The essence of belonging to the kind of humans heterozygous at the sickle-cell locus is simply being heterozygous at the sickle-cell locus.[3] Apart from obvious triviality, one reason why we should not take this seriously as a real essence is its failure to meet anything like the fifth condition. It does not constitute, in Locke's phrase, 'the very being of anything, whereby it is, what it is' (1975, iii. iii. 15, p. 417). For medical purposes it may be important to distinguish the class of people satisfying this condition, but the members of that class have countless and diverse other properties that sort them into countless other kinds. Other explanatory projects will be concerned with their membership of other kinds.

The fact that there are objectively correct ways of classifying organisms for particular purposes of enquiry, but that these classifications vary from one enquiry to the next, should lead us to prefer the weaker conception of natural kinds that I have suggested. As I shall now go on to argue, by considering examples from several central areas of biology, circumstances of the sort I have sketched for this case are typical of biology.

Systematics

Systematics, the study of biological diversity, is the obvious place to look for natural kinds in biology. Even quite recently philosophers have assumed that species, the principal units in terms of which biological diversity has been described, constituted natural kinds demarcated by real essences. But, as has been increasingly widely recognized, this is an untenable position. Most obviously, it is difficult or impossible to reconcile the idea that one species can evolve by gradual steps into another with the idea of unchanging and determinate

---

[3] The relevance of this remark should become clearer in the light of my discussion, below, of 'bare essences'.

essential properties. Reflection on the role of species in the theory of evolution has, indeed, led many philosophers and biologists to deny that species are kinds, natural or otherwise, and assert that they are a kind of temporally extended and spatially distributed individual.[4] The possibility that needs to be addressed here is whether 'species' itself might name a natural kind, either a higher-order kind of which the members were lower-level kinds, or a first-order kind whose members are species understood as individuals.

The first possibility can be dealt with quickly. It is doubtful, in fact, whether much sense can be made of the idea of a higher-order natural kind. Kinds do not have essences (if anything does, it is their members) and they most certainly do not figure as the subjects of laws of nature or possess causal properties.[5] One might perhaps suppose, to take the most promising sort of example, that 'halogen' named a higher-order natural kind, of which the members were the natural kinds chlorine, iodine, bromine, fluorine, and astatine. But the feasibility of this proposal will depend on the extent to which samples of the various halogens behave in similar ways simply by virtue of the common electron structure that make them all samples of halogens. If the halogens form a natural kind, that is to say, it is a first-order kind, whose members are the samples of any of the halogens. The biological parallel would be with the claim that genera, families, and such higher-level taxa were natural kinds. And this is widely agreed to be far less plausible even than that species themselves are natural kinds.

The appropriate parallel with chemistry would be with the claim that 'element' or 'compound' named natural kinds. But this is wholly unpromising. One knows nothing about the properties of a thing by knowing that it is a sample of an element, except perhaps, analytically, that it cannot be chemically decomposed. In parallel with an earlier suggestion about biology, it makes much more sense to think of these as concepts. Certainly they are very fruitful concepts, and

---

[4] Detailed elaborations of this view can be found in several essays in Hull (1989) and in Ghiselin (1997).

[5] On rereading this essay I was struck by an apparent tension between this remark and my discussion of the title question of the essay above (p. 105) and in the conclusion. I hope the tension is only apparent. What I have in mind here is whether there could be higher-level scientific kinds, whereas the other discussions cited consider whether natural kinds might form a metaphysical or methodological kind. The examples that follow here should make the appropriate distinction clearer.

many would hold that they are necessary concepts for any successful science of chemistry. But none of that tends to suggest that they are natural kinds. Species, in the present sense, is if possible a less promising candidate for the status of a natural kind than element. Nothing follows from the fact that something is a member of a species except, perhaps, that it is biological.

The prospects for treating species as a natural kind are clearly brighter if we assume that the members of the kind are individuals, if only because we are addressing an ontologically unproblematic candidate for a natural kind. The value of the proposal would seem to depend on two main factors. First, is there something common to all species that makes them a species, and second are there biological laws that apply to all species by virtue of their specieshood? I have argued elsewhere in some detail against the first claim on the grounds that there is no criterion of specieshood adequate to every domain of biological diversity (Chapters 3 and 4 in this volume), and I shall not repeat these arguments here. If I am right that 'species' when applied to birds, say, means something different from 'species' applied to bacteria, the second question can also be answered confidently in the negative.

Assuming for present purposes that a pluralistic interpretation of species is correct, an obvious possibility is that species encompass several distinct natural kinds. Perhaps biological species, in the sense of groups (or, more strictly, spatially scattered individuals) satisfying Ernst Mayr's Biological Species Concept, form a natural kind? Perhaps phylogenetic species of the kind theorized by cladists form another. I do not want to argue against these possibilities absolutely, though it is interesting to note that the various extant accounts of the species are typically referred to, like these ones, as *concepts*. What I would rather suggest is that they fit very uneasily with the traditional conception of a natural kind, but are easily accommodated by the much weaker view of natural kinds that I advocate.

Before getting into the details of this claim, I would like to mention an extremely heterodox context in terms of which the argument can be framed particularly clearly. For several reasons I think it would be very desirable to separate such putative natural kinds from the concept of a species. Outside professional systematics it is still generally assumed that the species is the basal unit of classification. It was natural to hope, in the aftermath of Darwin's theory, that the units of classification might turn out also to be units of evolution. But this attempt to connect species to evolutionary theory has been a failure.

Species concepts driven by evolutionary considerations proliferate, and no consensus appears to be in the offing as to which is most appropriate. Part of the reason for this is that different concepts seem much more appropriate to different groups of organisms. Although there is no dispute that evolution provides the ultimate explanation of biological diversity, it does not do so in a homogeneous way, and it does not do so in a way that reliably generates suitable units for general classificatory purposes. The solution to this problem, in my view, is to restore the term 'species' to its traditional home in classification, and allow theoretical units of classification to diversify independently as needed by specific technical concerns in biology.[6]

Reproductive barriers, the phenomena that the Biological Species Concept puts at the centre of the analysis of the species, are undoubtedly a fundamental cause of the generation and maintenance of diversity. The primary application of reproductive isolation, however, is not to the species but to the population. Species may consist of a single population, but they may also include numerous populations, sometimes wholly isolated from one another by geographical barriers or distance. The role of such isolation of populations in the process of speciation has been famously emphasized by Ernst Mayr (1963). But even Mayr admits that the question whether isolated populations are to be treated as belonging to the same species is one that can generally be settled only by inference from degree of morphological similarity (Mayr 1997: 130). Though complete isolation of sympatric populations may be taken as a sufficient condition of species difference, the reverse is by no means the case. The fact that sympatric populations exhibit significant exchange of genes is commonly taken as consistent with their belonging to different species. The only alternative to this concession is to admit that some species are massively polytypic and consequently of almost no use for classificatory purposes. Reproductive isolation is a matter of degree. In practice, groups of organisms are treated as reproductively isolated if barriers to genetic exchange are sufficient to maintain morphological difference between the groups judged to constitute specific difference. Concordance between the theoretical concept of reproductive isolation and a practically serviceable taxonomy is thus, in effect, maintained by linguistic fiat. The upshot of all this is that whereas the concept

---

[6] This is a very brief summary of an argument presented in detail in Ch. 4 in this volume.

of a reproductively isolated population is essential for understanding processes fundamental to the generation and maintenance of biological diversity, it certainly does not provide anything like an essential property of the species. And in view of the variation in degree of reproductive isolation observed throughout biology, it would be misleading to think even of the biologically isolated population as a natural kind. It is, rather, a property exhibited to a greater or lesser extent, with a variety of different causal bases, by many biological populations.

Closely parallel remarks can be made about the attempt to define species in terms of phylogenetic history. Although there is considerable disagreement about how this project is to be developed in detail, the basic idea is that a species should be monophyletic. On one interpretation this means that it should include all and only the descendants of some ancestral population.[7] Even if this is taken as a necessary condition for being a species, it is clearly far from sufficient. It may well be that all the mammals, say, are derived from a single ancestral population, and the definition would apply to a human family marooned on an isolated desert island. It would also apply to every clone of an asexual species. Some further criterion is clearly required for saying which such groups are the right size for a species. It is difficult to see any principled ground for this other than classificatory convenience. The most sensible response to this problem has been to insist on monophyly as a necessary condition for being a species, but allow a pragmatic pluralism when it comes to deciding which monophyletic groups should count as actual species (Mishler and Donoghue 1982). The problem then is that having effectively given up the possibility of seeing species as even a weak natural kind—they are an arbitrarily selected subgroup of the possibly natural kind monophyletic group—the grounds for insisting even on the necessary condition become quite unclear. One response to this problem has been to advocate the abandonment of the entire Linnaean hierarchy in favour of a system that would more faithfully express the phylogenetic position of any organism classified (Ereshefsky 1999; Mishler 1999).

---

[7] There is in fact debate over whether all the descendants need to be included, reflected in disagreements over the proper interpretation of the term 'monophyletic'. For purposes of the present discussion it will be sufficient to focus on the stronger, cladistic interpretation given in the main text.

But this seems to me a gross overreaction, and a disastrous one for those many consumers of the Linnaean hierarchy, from ecologists to mushroom collectors and landscape gardeners, for whom it is an excellent classificatory tool. Much better, surely, is to recognize that the group of organisms descended from a common ancestor is an important concept for phylogenetic analysis. Indeed it has a perfectly good name: the clade. Mapping evolutionary history is a fascinating and perhaps important project, but not so important as to be allowed to wreck a perfectly good classificatory system. Only the idea that it had finally been discovered what species really were, the essence that constituted them as forming a natural kind, could tempt one to pursue this iconoclastic path. But species do not form a natural kind, and there is no such thing to be discovered. 'Clade' perhaps names an important natural kind (in my weak sense) for phylogenetic analysis, but that is the most that it does.

The upshot of this discussion can now be quite simply summarized. Investigation of the processes of evolution draws our attention to more or less reproductively isolated populations, the recognition of which can perhaps be marked as that of a weakly natural kind. Recording the history of evolution, its product, points to coherent lineages of organisms, or more narrowly, clades, which can be similarly recognized. Neither of these is uniformly suited to the project of cataloguing biological diversity, so we would do well to distinguish the species as a classificatory unit which, if we are lucky, will coincide with one or both of the preceding. If this is right, however, the traditional view of natural kinds, leading us to expect an unequivocal answer to the question what is the unique real kind of which each organism is a member, will be disastrous. For it will lead us to assume that these three things must coincide. The oscillations between reproductive, genealogical, and phenetic solutions to the species problem perfectly record the attempts to square the circle from each of these perspectives.

Population Ecology

Population ecology is concerned with the determinants and dynamics of population numbers. It is not a science that has particularly encouraged essentialist aspirations. Still, it is a science in which the phenomena seem to dictate particular modes of classification, so it should be relevant to the issues with which this essay is concerned.

It is an analytic truth that changes in population numbers are determined by the difference between additions to the population and losses to the population. The processes that generate additions to population—reproduction, broadly construed to include various kinds of cloning—and losses—various causes of death—are well studied. Most organisms devote themselves as best they can to reproduction until death intervenes, so a study of determinants of death will generally be the key to understanding population changes. Most organisms, finally, die either because they are consumed by other organisms, or because they fail to find suitable organisms, dead or alive, to consume themselves.

This (banal, I hope) summary of population dynamics necessarily draws our attention to the relations of consumer and consumed in which organisms stand to one another. Hence we are led to think of organisms as predators and prey, parasites and saprophytes, and so on. And the models to be found in the toolkit of population ecology will generally include representations that refer to such kinds. In so far as the steps that lead us to such classifications are largely forced on us by the very simple empirical observations just mentioned, we may say that such categories provide natural kinds for ecology.

It hardly needs to be argued, on the other hand, that we have nothing like traditional natural kinds here. The most obvious reason for this is that most or all of these categories apply, from different perspectives, to any kind of organism. From the point of view of a worm, a thrush is a predator; from the point of view of a sparrowhawk it is prey. Even a tiger, which we might think of as a paradigmatic predator, is prey from the point of view of the fleas and lice that frequent its exterior and the worms that inhabit its interior. And on its demise, of course, a new set of consumers large and small will quickly take care of its remains. As these examples begin to suggest, and as is evident to anyone with the slightest familiarity with natural history, the ways of being, say, a predator are enormously diverse. At any rate, successful modelling in ecology is probably only possible to the extent that certain kinds of organisms have populations determined substantially either by a particular resource upon which they feed, or by a particular predator for which they are food. Since many organisms have extremely specialized ways of life, this is perhaps not an infrequent circumstance.

There is no a priori reason why the kinds appropriate for ecological modelling should invariably coincide with any of the kinds of kinds distinguished in my earlier discussion of systematics. From the perspective of determining the population of an avivorous hawk, it may be sufficient to estimate the population of small birds. From the viewpoint of the worm or snail, on the other hand, there is a difference of life and death between carnivorous small birds such as blackbirds and thrushes, and fruit-eating finches. A polymorphic species in which there is variation in foraging behaviour may, from an ecological point of view, need to be treated as constituting several distinct kinds. The point, once again, is not that there are no real divisions in nature between kinds of things, divisions that are appropriate for a particular kind of enquiry, but that what those divisions are will depend on what the enquiry is.[8]

Genetics

Those nostalgic for the ordered, crystalline, view of science that the traditional view of natural kinds supports may be tempted to look at the most modern and currently prominent area of biology, genetics. Succumbing to this temptation will rapidly lead to disappointment, however. The hope that genes may form a traditional natural kind founders rapidly on the discovery that no one seems to agree what a gene is.[9] We can all agree, perhaps, that terrestrial genes are composed of DNA or RNA. But exactly which sequences of DNA or RNA constitute genes is another matter.

Genes are most often heard of in contexts such as 'gene for blue eyes' or 'gene for intelligence'. We can be certain that these do not name natural kinds, probably not even in my weak sense. As has been continually emphasized in discussions of genetic reductionism, many distinct stretches of DNA are required for an organism to produce blue eyes, and almost all of those stretches will play a role in the production of other phenotypic features. There is no piece of DNA that is exclusively and solely the gene for blue eyes. Nonetheless, since most processes of natural selection select organisms with particular features, in so far as genetics plays a role in evolution, it appears to be in the form of such things as genes for blue eyes. If selection favours

---

[8] Related aspects of population ecology are discussed in more detail in Dupré (1993, ch. 5).

[9] See Kitcher (1982) for an early explanation of this point.

blue eyes, it will, *ceteris paribus*, favour genes for blue eyes. (This is, of course, not an endorsement of the view that selecting genes for this or that is the universally correct way of seeing selection processes that act on organisms.)

It is perhaps more promising to start with a bottom-up perspective and consider the possibility of distinguishing genes in terms of processes at the molecular level. The most familiar such process is that by which genes direct (or 'code for') the production of amino acid sequences. It is quite common to encounter 'gene' defined as the DNA sequence that codes for a distinct protein molecule. A surprising, though perhaps harmless, consequence of this definition is that most of the genetic material turns out not to consist of genes. More serious is the omission of bits of DNA that are surely functional but do not code for proteins. Best known among these are the so-called regulatory genes. Perhaps, again harmlessly, these form a distinct natural kind or natural kinds. Perhaps it is even harmless to conclude that the large amount of DNA that appears to be non-functional from the perspective of the organism does not consist of genes at all.

I am, of course, happy to conclude that genes in the sense being considered constitute a (weak) natural kind from the perspective of organismic ontogeny. What I want to insist on, however, is that they fail to satisfy most of the criteria distinguished as central to traditional natural kinds.[10] I have some concerns about the first such criterion, the determination of membership of the kind by an essence, but I shall postpone consideration of these to the next section. Here I want to consider rather some worries about the third and fourth conditions, about the role of the essential feature of members of the kind in explaining their characteristic behaviour.

Let us imagine, for the sake of argument, that all genes provide templates for producing proteins. Then certainly, given the universality of the genetic code, there will be an important property of any gene, the capacity to direct the production of a particular protein when situated in a functional biological cell, that is explained in terms of

---

[10] There is an obvious question whether we should take genes in general to form a natural kind, or particular genes (e.g. the human haemoglobin gene), or both, the latter being subkinds within the former. The first proposal raises obvious difficulties with relevance to laws of nature, in view of the great molecular heterogeneity of its members. The second suggests worryingly numerous and narrowly relevant kinds. Most of the difficulties I propose would apply to any of the interpretations listed. Otherwise it should be clear from the context which I have in mind.

the precise molecular structure of the particular gene. But, equally certainly, this does not exhaust the scientifically interesting properties of the gene. From either an evolutionary or a developmental perspective we may be interested in any of the range of phenotypic effects to which the gene may contribute. Suppose we are interested in the class of genes for blue eyes. This kind will include all of those genes which, in an appropriate genetic and extra-genetic context, will tend to produce blue-eyed phenotypes. Since the particular gene that has this tendency will typically have other effects, for example to produce light-coloured skin, it will also belong to a kind, gene for light-coloured skin, that will typically only partially overlap with the first kind. Thus the molecularly classified gene will also belong to a range of disjoint kinds when classified from this phenotypic perspective.

Moreover, it cannot be assumed that every gene satisfying the molecular description will be a member of the class of genes for blue eyes. It is now well known that precisely the same molecular gene often occurs in countless different species. The gene in question will very probably occur in species of which no members have blue eyes. (Perhaps they have no eyes at all.) Those occurrences of the gene can certainly not be classified as genes for blue eyes, since they have no tendency to produce this phenotypic feature. So, once again, different investigative concerns will lead us to classify together quite distinct entities. So, again, we cannot give a unique answer to the question what kind of entity this gene essentially is. Things belong to kinds only from a particular perspective and so are, at best, only natural kinds in the weak enquiry-relative sense I have been advocating.

Bare Essences versus Hidden Essences

The preceding argument has rejected a traditional essentialist view of natural kinds for several central areas of biology. I have not, however, advocated extreme nominalism. In all the cases I have considered I have suggested that the development of an enquiry with particular goals has led naturally to the discrimination of particular kinds, kinds which for that reason I am inclined to call 'natural'. And it may perhaps have often seemed plausible to suggest that, within this particular context of enquiry, the kinds in question are determined by an essential property. From the point of view of a phylogenetic taxonomy, for instance, being descended from a particular ancestral population may seem to be the essential property of a particular taxon. From the point of view of the population dynamics of birds, the disposition to consume small birds may look like the essential property of predators. And from the point of view of molecular genetics, possessing a particular sequence of base pairs may seem to constitute the essence of a particular gene. So perhaps the argument should be taken to show just that there are more natural kinds than has sometimes been thought. And perhaps the only aspect of traditional essentialism that we need to abandon is the idea that the essence of a thing provides an unequivocal answer to the question what kind of thing it is. Perhaps, finally, things should be said to have a number of essences, each of which explains a reasonably well-defined aspect of the behaviour of the thing.

I am generally sympathetic to the preceding suggestions, and indeed they come close to summarizing the view that I have in the past referred to, and defended, as promiscuous realism. However, whereas I have advocated in this paper and elsewhere co-opting the expression 'natural kind' for such a picture, I am less happy to co-opt the term 'essence'. This is partly just because saying that a thing belongs to a variety of natural kinds seems to violate traditional usage much less dramatically than to claim that a thing may have several essences. But there is a further difference that such a co-optation of usage would tend to obscure.

The traditional conception of essence sees the essence as a goal of enquiry. Enquiry begins with an antecedently recognized kind, and seeks to understand its nature by discovering the essential property that makes a thing a member of that kind. This is as clear in the recent essentialism of Kripke and (early) Putnam as it is in the ancient essentialism of Aristotle. Essences are the hidden reality underlying and explaining the existence of a kind. This picture wholly misrepresents the role in enquiry of the kinds of kinds I have discussed in this essay. In real scientific enquiry the recognition of the kind is identical to the formulation of the 'essential property'. This is why I suggest calling such essences 'bare essences'. Once the kind is distinguished, the essences lie open to view. It seems to me precisely the failure to make this distinction that underlies so much futile argument on the species problem. The discussion of species still generally proceeds as if we know perfectly well that there are species out there, and we know how to distinguish many of them; it is just very hard to discover what they (essentially) are. But this, I maintain, is to formulate an unanswerable question which has, unsurprisingly, proved difficult to answer. Rather, there are various questions connected to biological diversity. What are the coherent units of evolutionary change? What are the phylogenetic relations between different kinds of organisms? How can we best communicate and record information about organisms? In addressing these questions we generate various principles for classifying organisms. It would be nice if at the end of these processes we came out with identical sets of classifications which we could then triumphantly declare to be the true species. But the evidence suggests that this is an improbable outcome. And the assumption that satisfactory progress with addressing any of these questions is evidence that our principles of classification have captured the essence of the species is a relic of an antique essentialism that we should surely reject.

A final approach to the general message of this section is to recall Mayr's (1997) emphasis on the role of new concepts in biological progress. As I suggested earlier in this essay, deciding how to sort the phenomena within a domain into kinds is much the same thing as deciding what concepts are to be applied in classifying the relevant domain. This suggests again that we are dealing with kinds distinguished by a purely nominal essence: to be a member of the kind is just to satisfy the concept that was used in distinguishing the kind. There is nothing wrong with this, in the end. The only problem is that it is liable to obscure the fact that discovering what concepts to apply is at the heart of successful research. It is a product of successful enquiry that we should apply the concepts heterozygote and homozygote in understanding the persistence of certain apparently deleterious genetic conditions, for instance. There is a historical tendency to associate talk of nominal essences with arbitrariness and triviality. This is perhaps quite generally mistaken. The process of deciding what concepts to use, whether in science or in everyday life, is perhaps best seen as embodying the collected empirical experience of the human race. Still, I think that an appropriately weakened notion of kind will serve to fend off inappropriate connotations that might be encouraged by an analysis solely in terms of concepts and correlative nominal essences.

Conclusion: How Natural are Natural Kinds?

I have suggested throughout this essay that although scientific classifications do not accord with the traditional, essentialist conception of natural kinds, they do, nevertheless, represent objectively appropriate ways of classifying in the light of particular investigative ends. This brings to mind a question that has been raised in recent work by Ian Hacking (e.g. Hacking 1999) whether the outcome of (successful) scientific enquiry is somehow determined in advance by how things objectively are. This is not at all, incidentally, the same thing as a question about scientific truth, or realism. There are, I suppose, infinitely many truths, and the truths of science may be doomed to remain an infinitesimal proportion of all the truths. Nevertheless, it is often suggested that certain truths, such as Maxwell's equations, or Newton's laws, or the periodic table of the elements, would be a part of any successful science.

The degree to which I want to insist that scientific kinds are, indeed, natural clearly contradicts the most extreme nominalism which, by taking classification to be entirely unconstrained by nature, implies that the end products of research (theories, laws, models, etc.) must be equally unconstrained. On the other hand, the denial that there is a uniquely correct way of classification leaves it open whether any particular such end products are unavoidable. The correct resolution of this difficulty, I suggest, is that whereas sharply formulated questions within a specific context of enquiry may generally determine the form of an adequate set of answers, the sets of questions that are in fact formulated is a matter of considerable historical contingency. (This is, as I understand him, close to the view that Hacking advocates.) It is plausible that there is an important path dependency to scientific knowledge. Solutions to one set of questions make certain future developments salient, others more or less unreachable. This does not preclude the possibility that certain elements of theory may be more or less inescapable. If we conceive of science as following a number of paths through an infinite erotetic space, it may be that, like Rome, all paths lead to, or pass through, the atomic theory, say; or at least that so many paths do that this theory is almost impossible to avoid. But here we are in the realm of questions that can only be sensibly addressed by detailed attention to the history of science, attention that I certainly do not propose to offer here.

What I mainly want to emphasize in these concluding remarks is that large questions about the general nature of science can illuminatingly be approached through consideration of the nature of classification. In particular, what I take to be the correct view of classification suggests a promising middle way through the very heated contemporary controversies about scientific realism and social constructivism. Traditional essentialism leaves little space for the insights into science that have been provided by the best historical and sociological work carried out under the constructivist banner. This has led many philosophers of science to reject the possibility of any such insights and insist on the absolute and unique objectivity of scientific knowledge. The pluralistic picture I have been sketching, on the other hand, allows that factors of many kinds may be needed to explain the path that a scientific research programme follows through erotetic space. But it also leaves open the possibility that, from the perspective of a particular point in that space, there may be a uniquely optimal way to understand the phenomena with which science is at that point confronted. By the identification of natural kinds I mean to mark, more or less, the experience of recognizing the taxonomic aspect of such an optimal understanding. Whether this is sufficiently related to traditional conceptions of natural kinds to justify the continued use of the term strikes me as relatively unimportant.

There is an obvious proviso that I should enter here. This paper has been concerned exclusively with various areas of the biological sciences. It would be wholly inconsistent with my pluralistic attitude to assume that similar results would be found in other areas of science. But of course pluralism has the great advantage over monism that it is established by one instance (or anyway two). So even if the physical sciences were best understood as requiring a single unique and necessary scheme of classification, science as a whole will still have been shown to be plural in nature. (Unless, of course, as some philosophers have been tempted to assert, the only sciences are the physical ones. That is, at any rate, a bit of linguistic legislation that I am unwilling to adopt.)

Let me finally return to the question of my title. It seems to me that 'natural kind', understood in the sense I have proposed, might be seen as a natural kind of methodology, or of the philosophy of science. It marks the view just summarized that sufficiently well-formulated research projects will often lead to what appears to be an optimal, even an unavoidable, set of classifications. On the other hand, it would be wrong to think of natural kinds as a metaphysical natural kind. Essences, the hidden inner constitution of things that make them members of the kinds to which they belong, and explain their behaviour in accordance with laws that apply to the members of such kinds, would, perhaps, be metaphysical natural kinds. But I do not believe there are such essences, certainly not in biology. The classifications I have discussed in this essay are grounded in factors as diverse as physical structure, historical origin, role in a system of interdependent entities, and so on. These share nothing but the bare methodological essence of playing the important roles they do in the theories that they inform. So my weak natural kinds are at best a weak natural kind for the philosophy of science, surely not a traditional natural kind. Traditional natural kinds would, perhaps, form a traditional natural kind. But even if there is such a kind, they do not form even a weak natural kind from the perspective of the philosophy of science. Conceiving of science in terms of such a kind leads only to error and confusion.

IV
Kinds of People

6

Human Kinds

Introduction

A critical dimension of debate about any large-scale scientific theory is the question of the scope of its generalizations. It is this issue, with respect to the application of evolutionary theory to behaviour, particularly human, that I want to address in this paper.

Many very disparate academic disciplines claim relevance to the understanding of human behaviour, and these disciplines vary very widely in their pretensions to scope. At one extreme are the main humanistic disciplines (though excluding most of philosophy). Historians may or may not imagine universal processes or laws emerging from the narratives they construct, but they have not primarily achieved fame for the vindication of such generalizations. Literary critics may have general theories about the social functions of texts, but their main professional activity is the study of the most idiosyncratic of cultural artefacts. Among the social sciences, cultural and symbolic anthropology, and much of sociology, is equally focused on such cultural specifics. In such disciplines, discussion that is not so concretely focused almost always concerns methodology rather than the postulation of universal laws.

Uncontroversially at the other extreme are the medical sciences, especially those concerned with behavioural pathology, and a large part of psychology. Since only the most ignorant of racists deny that *Homo sapiens* is a respectable biological species, it is to be expected that strictly physiologically based studies should claim species—or often wider—scope. The psychology of perception, or much of learning theory, can clearly be conceived of in this way. There is no reason to doubt that there are commonalities among humans about the way information is collected and processed. Shepard (1987) suggests ways in which evolutionary ideas may be applied to such very general aspects of the mind.

This paper has benefited greatly from some extremely thoughtful and pertinent comments on an earlier version by Peter Richerson. I am also very grateful to the Stanford Humanities Center, where this paper was written.

Studies of these two kinds generally avoid conflict with one another. This is because the studies with the broad scope do not generally claim to explain behaviour in anything like the detailed way possible for theories of the first kind. The mechanisms of perception, for instance, while obviously necessary for much human behaviour, would hardly be expected to determine very much behaviour on their own. Relative to the explanation of behaviour, then, we might distinguish between those disciplines that aim at breadth but not depth, and those that sacrifice breadth for depth.

Conflict is likely to arise when a theory attempts to claim both breadth and depth. Economics has sometimes, though generally cautiously, tended to make claims of both these kinds. Forthright and unambiguous is E. O. Wilson's notorious manifesto at the beginning of *Sociobiology* (Wilson 1975). Less aggressively, sociobiologically inclined anthropologists have argued that the potentially universal scope of evolutionary arguments gives an evolutionary methodology a fundamental synthetic role linking the more parochial approaches traditional in the social sciences with each other and with biology.[1]

I want to approach the question of the scope and depth of evolutionary explanations of human behaviour by way of an analogy with the explanation of non-human behaviour. As a partly heuristic device, it will be useful for this purpose to introduce the notion of a 'cultural species'. I do not mean this concept to be taken wholly seriously, though it does effectively dramatize a major part of the argument I want to present. At the end of the paper I shall indicate one point at which, I believe, the analogy implicit in it breaks down.

Culture and Human Behaviour

The attempt to apply evolutionary ideas to human behaviour has constantly elicited the protest that any such project must disregard the centrality of culture to human life, and is either hopelessly naive or even malicious. Evolutionists retort that human culture is merely an admittedly unique biological phenomenon that appeared at a particular point in the evolution of a particular species of primate, and should be amenable to the same general treatment as any other

---

[1] See Smith (1987) for such a claim and further references; a philosopher with a similar, but even stronger, position is Alexander Rosenberg (1980).

biological phenomenon. In particular, it is argued, it could not have evolved if it had not been adaptive. And so, subject to whatever limitations may be recognized in the general assumption of optimal adaptation, human behaviour, whether culturally or genetically transmitted, should be subject to the standard kind of evolutionary analysis. Some account of the source of the allegedly independent causal force of culture seems at least a reasonable demand. To a considerable extent, participants in this debate, for want of any agreement about adequate standards of explanation, have simply argued at cross purposes to one another.

The chances of genuine progress have been greatly augmented by recent theoretical studies of the possibility of autonomous[2] processes of cultural evolution (Cavalli-Sforza and Feldman 1981; Boyd and Richerson 1985). Such studies start from the observation that culture might evolve in a way closely analogous to, but also quite distinct from, standard Darwinian evolution. It is generally agreed that an evolutionary process will occur whenever a class of organisms displays heritable variation in fitness. It is highly plausible that cultural variation should satisfy such a condition. On the other hand, the details of such a process may be very different from those of the familiar Darwinian one. First, the transmission of culture occurs through different channels; there is no reason to assume that it will primarily, or anyway solely, occur from biological parents to offspring, and it may occur laterally, within rather than across generations. Second, it will presumably be Lamarckian; one would expect acquired cultural variation to be transmitted. And third, there is no reason to assume that the appropriate concept of fitness will coincide with the equivalent Darwinian notion. Presumably the appropriate concept of fitness will depend on some measure of the tendency of different cultural variants to be transmitted to other individuals. Though arguments have been proposed purporting to show that, at least in the long term, the two concepts of fitness will tend to coincide (Durham 1978, 1981), Boyd and Richerson show in considerable detail why this may not be the case.

---

[2] Peter Richerson (personal communication) has expressed some concern about my use of the word 'autonomous', pointing out that he and Boyd model cultural evolution as a process intimately linked to genetic evolution. I certainly do not mean to take a stand against this claim. What I do mean to stress is that, first, culturally driven processes have major weight as partial contributors to the course of human evolution, and second, for many aspects of human evolution they may be the most significant forces at work.

The great achievement of this work is that it shows precisely what kinds of processes can lead to an independent role for culture in the determination and explanation of behaviour. Rather than blankly confronting evolutionists, or sociobiologists, with incompatible ways of approaching human behaviour, and asserting their superiority, it confronts them precisely on their own ground. Culture is shown to be fully amenable to integration within a broad evolutionary perspective, and also to have a potentially important role in evolution. On the other hand, while Boyd and Richerson are prepared to argue that some autonomous processes of cultural evolution have been significant in human evolution and offer some specific if tentative instances, they are generally agnostic about the relative importance of genetic transmission and cultural transmission in the determination of human behaviour, and leave open the possibility that genetically determined biases with respect to the transmission of culture may have led cultural evolution predominantly to converge on the results that would be predicted by more conventional sociobiology. In this paper I want to argue that culture variation, very probably the product of divergent cultural evolution, should be seen as the primary focus for the explanation of human behaviour.

Cultural Species

It will be useful to introduce this argument by recalling a fundamental explanandum of the Darwinian theory of evolution, the diversity of life, or, in a well-known phrase, the origin of species. Note that the diversity in question is that between, not within, species. The latter is an essential part of the explanation of the former. Of course, the details of the process of speciation remain highly controversial to this day. But that it is a primary goal of the theory of evolution to explain it, and that no such explanation would be possible without the existence of intraspecific variation, is not in dispute. And this indicates a striking disanalogy with the theory of cultural evolution as it has mainly been developed. As so far described, the theory is largely concerned with anagenetic evolution within one lineage, *Homo sapiens*.[3] My

---

[3] The apparent rarity of cultural transmission in non-human species is briefly discussed by Boyd and Richerson (1985: 130–1); see also Pulliam and Dunford (1980). I shall not be concerned here with this fascinating and important question.

suggestion is that the full import of the theoretical development only becomes apparent when human evolution is considered as a clado-genetic process involving the generation of many different (cultural) species. It is to explore this suggestion that I want to consider the significance of treating distinct cultures as genuine taxonomic entities, what I am calling 'cultural species'.[4]

The importance of this perspective, however, derives not so much from the role of evolution in explaining diversity as from the role of taxonomy, or the description of diversity, in evolutionary explana-tions. This role may best be understood by adverting to another fundamental issue in evolutionary theory, the problem of holism. A central difficulty in adaptationist methodology is the existence of crucial limitations on the atomistic approach generally characteristic of this methodology. These limitations have been identified at various levels of organization. Beginning at the genetic level, probably the least serious difficulty is the phenomenon of genetic linkage, the fact that Mendelian independent assortment breaks down when genes are close together on a chromosome. More significant is pleiotropy, the observation that genes may affect a variety of apparently quite distinct phenotypic properties. And perhaps most important is the phenom-enon of epistasis, the dependence of the expression of genes on the action of genes at distinct loci. While these phenomena have mainly been emphasized as showing the limitations of models of single locus selection, they are symptomatic of a much more fundamental fact, the enormous degree of integration of a biological organism.[5] What I want to emphasize here is that the way that a particular kind of organism is integrated, the way its parts and characteristic forms of behaviour fit together, while certainly a product of past evolution, is also the most fundamental constraint on the evolutionary process. Thus to whatever extent general laws can be formulated about the course of evolution, their consequences cannot be understood in any concrete way without an account of the nature, past and present, of

---

[4] A similar suggestion can be found in Lorenz (1977, ch. 9). A specific application of such an attempt to the Indian caste system has been attempted by Gadgil and Malhotra (1983), though they concentrate almost exclusively on ecological factors. (I am grateful to Peter Richerson for this reference.)

[5] I take it that this is a central point of the much discussed critique of adaptionism by Gould and Lewontin (1979). Peter Richerson has pointed out to me that there is no neces-sary connection between the complexity of genetic interactions and the integration of organisms at the macroscopic level. Both phenomena, however, will give rise to analogous constraints on evolutionary possibility.

the particular lineage to which they are to be applied. And this kind of account can only be derived from a descriptive taxonomy of the organisms concerned. The elaboration of this claim, and its relevance to the study of both human and other animal behaviour, will be the central task of this essay.

Animal Behaviour and Taxonomy

To defend the centrality of cultural taxonomy to the explanation of human behaviour, I want now to develop an analogy with a case of non-human behaviour. Consider the question of understanding the behaviour of a fairly diverse group of animals, say the mammals, relevant to the acquisition of food. The question I want to ask is to what extent such explanations can be universal, and to what extent they must rest on quite specific aspects of the physiology and ecology of the particular species.

One obvious aspect of such an explanation that will be quite universal will be the most general function attributed to it. Animals seek food because they need energy for survival, reproduction, and whatever other activities they engage in. Almost equally universal, at least in general outline, are the physiological processes leading from the capture of food to the utilization of energy from that food. All mammals have mouths into which items of food are placed, most have some kind of teeth with which these items are physically degraded, all have stomachs and intestines in which the items are chemically degraded and absorbed, and all have anuses from which the unused components are excreted. Of course the exact processes of physical and chemical degradation are significantly differentiated, but I shall ignore those here.

Much more promising attempts at general explanation with prospects for addressing the specifics of behaviour are provided by evolutionary optimality theorists—in particular, by what has become known as 'optimal foraging theory' (Charnov 1976). This theory addresses questions such as how long an animal will persist in a particular episode of foraging, how many different kinds of food item it will seek, and whether it will share pieces of food with other conspecifics. My interest here is not in the adequacy of such analyses (though see Emlen 1987), but in what is involved in their application. And what is clear is that even if the theory were to apply perfectly to every possible case, no predictions about behaviour could be made without considering the details of the ecology and physiology of the type of organism concerned.

To begin with, it is a clear feature of such models that the optimal searching behaviour is a function of characteristics of the kind of food being sought. If the food consists of small items concentrated in particular areas, such models can give quite precise predictions, as a function of the density of food in these patches, about when the animal will abandon one patch and move on to look for another. On the other hand, if the food is uniformly distributed in small portions, or randomly distributed in large bundles, this particular type of model will have no application. Thus, however successful the model is in describing the decision procedure of an anteater deciding to move on to the next termite mound, it will tell us nothing about a horse grazing in a cultivated field or a tiger stalking through the forest.

Similarly, there is a very elegant analysis of what variety of food items a particular animal will be prepared to eat. And while this may be extremely illuminating in explaining why a shrew eats beetles but shows no interest in spiders, say, it will clearly have no application to anteaters—for the obvious reason that they are entirely specialized at the eating of termites. Moreover, even when we know that the animal is an omnivore, or a generalized carnivore, unless we know a good deal about its physiology, we are still in no position to draw any behavioural conclusions from such a model. If we knew merely that shrews were generalized carnivores, we might well predict that they would eat nothing but elephants. Slightly less absurdly, we might anticipate that grizzly bears should live mainly on deer or other large ungulates if we did not find out that they could not run fast enough to catch them.

A further point, emphasized by Emlen (1987), is that the way an animal looks for food may very well depend also on what other types of behaviour it engages in. A very general example might be the question whether the animal is nocturnal or diurnal. The prey species selected by a nocturnal carnivore that hunts at waterholes will evidently differ from those selected by a diurnal carnivore. A nocturnal way of life will require a whole complex of adaptations, such as suitable sensory equipment and mode of locomotion, which may well provide constraints on the sources of nutrition available to it. And of course there is no reason to assume that the method of foraging will be the decisive determinant in the evolutionary selection of a nocturnal life.

The point that is perhaps being laboured here is just that an atomistic treatment at a quite abstract level of one general aspect of behaviour generates no behavioural conclusions whatever. Even if the optimality analysis proved to be exceptionlessly true, it would have no application to a particular species without detailed understanding of how the possible modes of acquisition of food are integrated with the many other behavioural and physiological peculiarities of the species in question. The question that is then crucial for considering the human analogy is the following. How far must one pursue a process of taxonomic analysis before one arrives at a sufficiently homogeneous group of organisms for a meaningful application of abstractly conceived optimality analyses? And the rhetorical point of introducing the notion of 'cultural species' is to argue that for *Homo sapiens* this process must be pursued to a much finer level than merely the biological species.

Human Behaviour and Cultural Taxonomy

I want now to show why the application of such general adaptationist arguments to humans not only faces the same sorts of limitations as would apply to a general account of mammalian foraging behaviour, but does so in an even more severe form. And the problem is more severe primarily because even if it is defensible to consider the behaviour of non-human animals as involving the reconciliation of a very small number of goals derived from the general postulate of fitness maximization—feeding, reproduction, predator avoidance, perhaps—this does not seem to be at all plausible for the human case. I put this more cautiously than I would wish, because a major tendency in sociobiology, supported by sympathetic anthropologists and some economists, is committed to just this prima facie implausible thesis. I shall argue, however, that it is not only implausible, but also false. And consequently, the variation in the context into which a particular aspect of behaviour has to be integrated may actually be greater within the human species than even that across somewhat related animal species.

Let us again focus on behaviour involved in the acquisition of food, and again begin by asking what aspects of this behaviour may fairly be treated as human universals. Clearly, the ultimate, though by no means exclusive, function of such behaviour is common to all humans, and indeed is the same as in other animals. Equally clearly, the physiological context is much more homogeneous than that considered in the previous case, and this may reasonably be extended to include general features of the neurological processing involved. Perhaps some of these general features may be amenable to the kind of evolutionary approach suggested by Cosmides and Tooby (1987).

Leaving aside for the moment that a large proportion of the world's human population 'forages' by carrying pieces of paper of purely conventional significance to the supermarket or village store, let us focus only on the hunter-gatherer societies to which evolutionary analysis seems most readily applicable. Despite the physiological homogeneity, it is obvious that the geographic variability of human populations will be sufficient to require one crucial environmental constraint. One cannot select the same items of food in the Arctic as in a tropical forest or in a desert; so, no behavioural predictions follow from the general analysis, at least by itself. This problem may be considered relatively superficial, analogous to the common observation that no predictions follow merely from Newton's laws without detailed specification of the initial conditions to which they are to be applied. But even if we take a description of the environment as providing the necessary initial conditions for application of optimality analyses, there are much more serious obstacles to be faced. And these again involve the integration of one aspect of behaviour into a much broader pattern.

One basic difficulty that emerges here may best be expressed as a difficulty in determining what is being optimized. The great strength of optimality theory in biology is that for all the doubts about whether optimal solutions will in fact be achieved, or are even wholly determinable, it is roughly clear what, if anything, is being optimized. In the standard Darwinian conception, even when there are conflicting objectives, so that one goal must be treated as a constraint on the behaviour appropriate for achieving the other, at least a common currency exists for evaluating such trade-offs, namely, inclusive fitness. That such a resource is available in analysing human societies is much less clear. The problem may be analysed in two stages. First, ecological analysis of human behaviour will generally begin with assessing not fitness, but something treated as a promising proxy. Divergence of the proxy from actual fitness consequences has been recognized by ecological anthropologists, and causes great complications. But second, and ultimately more serious, other facts about the social organization of a society may provide constraints that are not even analysable into competing determinants of fitness. (The analysis in the first part of Lewontin 1987 suggests that problems of the kind indicated here may become unmanageable even apart from the special complexities in the human case.)

The difficulties of the first kind are fairly evident in the literature on optimal foraging theory. The simplest assumption is that hunter-gatherers should adopt a foraging strategy that maximizes the rate of energy acquisition. Presumably the adoption of such a strategy is to be seen as an adaptation, achieved either by group or by individual selection, with an undetermined mix of genetic and cultural inheritance (Durham 1981: 228–9). Since even 'primitive' peoples have many goals other than the avoidance of starvation, and foraging strategies may have relevance to some of these other goals, the plausibility of assuming that energy intake maximization will be a successful predictor of foraging strategy must depend on the assumption that energy acquisition is a dominant constraint on the survival or reproduction of either individuals or groups. But even apart from any other goals that may be directly served by particular foraging techniques, average rate of energy intake may well not be the appropriate variable to consider. If energy shortages are a recurring problem for hunter-gatherer groups (Winterhalder 1981: 67), it may well be that minimizing the variance of energy acquisition rather than maximizing the mean rate will be crucial (Smith 1981). Again, energy may not be the crucial constraint on the success of foraging. Linear programming models analysing the various goods and dietary constituents that can be important consequences of foraging suggest that much more specific factors may be important at different times (Keene 1981). While these complications do not show that this approach to anthropological explanation is hopeless, they do illustrate well the difficulties encountered by optimization in the absence of the clear definition of currency provided in evolutionary theory (difficulties, of course, in addition to the limitations of optimization within evolutionary theory).

The preceding difficulties show that human foraging is a very much more complex activity than, say, feeding gasoline into a car. But the fundamental difficulty is not so much this complexity but, again, the question of integration. Durham (1978, 1981) has argued that whether the determinants of optimal foraging behaviour are cultural or genetic, ultimately selection within and between groups will guarantee that this optimum is approached. Presumably this would be true if there were infinite variation and sufficient selection. But a general point lying behind the subordination of optimality to taxonomic fact is that there is far from infinite variation. The available options on which selection can work are both highly restricted and—which has been my main point so far—highly integrated. In both the human and the general cases, the integration, by presenting only a relatively small number of structured packages of features, is a primary explanation of the limitations on evolutionary possibility. The force of the analogy I have been trying to develop is that the integration of organisms with a specific physiology and a specific behavioural repertoire quite generally provides an extremely powerful set of constraints on the optimal adaptation of a particular domain of behaviour. The further point in considering human behaviour is that specific cultures are also highly structured systems. My purpose in considering the notion of cultural species is to emphasize that this is a level of organization quite as significant as the specific adaptation of an animal species as a constraint on behaviour.

I do not mean to imply that ecological anthropologists typically ignore constraints of these kinds. On the contrary, and in contrast at least to what Philip Kitcher (1985) has called 'pop sociobiology', most evolutionary anthropologists aim only to describe quite local fitness maxima. My worry, however, is not merely that local circumstances will determine different fitness maxima, but rather that aspects of the overall social integration of human groups may constrain behaviour in such a way that no optimum in terms of any remotely simple features even of the local environment will be attainable.

Let me illustrate the kind of thing I have in mind with two crude examples. Suppose that a certain species of lobster proves to be one of the most efficient sources of energy and nutrients in the environment of a particular tribe. However, age-old tradition has it that only the aristocracy is permitted to eat lobsters (as, for instance, in England the consumption of swans is the prerogative of the royal family and the fellows of St John's College, Cambridge). Hence lobsters, from an optimal point of view, are greatly underutilized. Or perhaps monkeys would be an ideal food source, but the local religion considers them unclean (miscreants are supposed to be reincarnated as monkeys, for example). Historical experience does not suggest any overwhelming tendency for such arbitrary conventions to collapse in the cause of efficiency; and there may well be strong functional reasons, again to be understood in terms of the particular articulation of a particular culture, why they are maintained.

Assume, also, that the tribes in question get by quite well without eating lobsters or monkeys, though perhaps with a slightly lower equilibrium population. The point about infinite variation should now be clear enough. There may be no other groups competing in the same territory; and if they are, they are not likely to differ only in respect to whether lobsters or monkeys are part of the diet. The determinants of competitive success may be quite unrelated to diet—the monkey shunners, for instance, have invented the bow and arrow. There is no relevant individual selection, since anyone found illegally eating a lobster is summarily executed.

To summarize the preceding argument, my main thesis is that very little in the way of a behavioural conclusion can come from a general evolutionary argument. This is not an argument against the legitimacy of abstract optimality arguments in evolutionary theory, but a point about their place in a general theory of behaviour. The analogy mentioned above, between evolutionary theory and Newtonian mechanics, indicates a starting point for my contention: general theories have no consequences without specifying initial conditions or constraints. The situation in evolutionary theory is worse, however, because an abstract optimality argument can only indicate one among many possible evolutionary forces (see Richerson and Boyd 1987; Sober 1987). As the complexity of an organism increases, so will the number of relevant forces that may come into play. Not only do we need a detailed description of the particular organism to understand what the significance of a particular force, *ceteris absentibus*, would be, but we need such information to understand how various such forces have interacted in the organism's evolutionary history, and now determine its mode of existence in its current environment. The analogy with mechanics would begin to be appropriate only if we add to the determination of initial conditions the stipulation that the bodies concerned were additionally endowed with various degrees of electric charge and magnetism (the point thus parallels part of Nancy Cartwright's (1983, essay 3) argument for the falsity of the laws of physics).

The specific claim about humans is that part of the information we need is the detailed environmental parameters of a particular human being. I am suggesting that cultural evolution, in relation to this explanatory problem, has provided variation quite comparable to the physiological variation generated by Darwinian evolution; and features of the cultural environment that exhibit this variation are indeed among the decisive determinants of human behaviour. Moreover, there is little basis for the assumption that these determinants can even approximately be reduced to genetic fitness. I have tried to indicate some general considerations that show why this is so; Boyd and Richerson have examined some specific processes with this consequence in a much more rigorous way. These points go a long way towards justifying the appeal to the notion of a cultural species.

The point of all this is not, of course, that *Homo sapiens* is in any way an inadequate *biological* species. The question is rather whether the biological species category is an adequate one for analysing human behaviour. My answer, evidently, is in the negative. The claim that I have been trying to defend is only that cultural distinctions within the (biological) species should play a role in our understanding analogous to that played by distinctions between species in biology generally. Obviously this is not an attempt to identify the one concept with the other. It would thus be glaringly irrelevant to point out that there is substantial gene exchange—or even, for that matter, cultural exchange—between human groups. (I do not know how the pattern even of *gene* interchange between human cultural groups would compare with, say, that between various species of Compositae or Rosaceae. But as I have just said, this question is of no relevance to the present discussion.)

Let me finally put the issue in a rather broader philosophical context. Biologists (such as Ernst Mayr 1963) and philosophers of biology (such as David Hull 1965) have been arguing for many years against essentialist thinking in biology. One aspect of essentialist thinking that is particularly difficult to eradicate is the idea that *if* one has identified the fundamental taxonomic entity to which an object is to be assigned, then it is by reference to properties characteristic of that class that ultimate explanations of the behaviour of the object must be constructed. But first, there is no reason to believe that there is such a thing as a 'fundamental taxonomic entity'. And even if there were, good empiricism requires that we in no way prejudge the issue of how much of the behaviour of the entities belonging to such a class is uniform within it. (I have discussed the latter point at some length elsewhere; Chapter 8 in this volume.) Thus the undoubted fact that the species is the fundamental level of distinction among biological organisms[6] (arguably the controversy about

---

[6] This assumption has become more controversial since this chapter was written. See e.g. Ereshefsky (1999) and Mishler (1999).

how species should be defined reflects the fact that this is taken as true by definition) tells us nothing about what range of generalizations will hold within a particular species. Thus there is at least no conceivable a priori objection to my claim that within the particular biological species *Homo sapiens* behavioural generalizations usually fail. For this purpose a finer taxonomy is essential. The claim that humankind encompasses many human kinds is, I believe, a perspicuous statement of the objection that has so often been entered against human sociobiology on behalf of the importance of culture.

Cultural Species and Forces

I have been arguing that it is unprofitable to approach the study of human behaviour with a search for universal principles. Though there will certainly be rough universals about such matters as the general ways in which humans process information, or acquire language, specific dispositions to behaviour will be dependent on a much more detailed description of the cultural context in which the particular human is situated.

I have also suggested that cultural diversity should be understood in many respects in the same way as biological diversity—that is, as the result of an evolutionary process. In particular, individual variation, and processes analogous to selection, should be seen as providing a historical basis for the present existence of various internally articulated and integrated cultural forms. I shall also make some tentative suggestions later about one important possible source of such variation. But first, I want to emphasize in another way the appropriateness of this way of thinking about these issues.

Part of the significance of an adequate taxonomy for understanding biology might be put in the following way: the species to which an organism belongs will be an excellent and indispensable predictor of its susceptibility to the various environmental forces that impinge on it. For example, the appearance or introduction of a certain parasite might have devastating effects, including even extinction, on certain species, while having none at all on most others. A drought might have very little effect on suitably adapted plants, while being fatal to whole species that are adapted to higher water consumption. I shall suggest, analogously, that the major forces that have an impact on humans are cultural rather than biological, and that their impact is similarly susceptible to the cultural contexts of individuals.

Consider, for instance, a very biological-looking force, a food shortage. In some societies it might happen that this would provide selection for the capacities to produce food efficiently: hunting, gathering, raising produce, etc. But clearly this is not at all the impact of such an occurrence on most of the societies with which we are familiar. Consider the great Irish famines of the nineteenth century. Presumably the Irish peasants were actually more skilful in the production of food than the English landowners (mainly noted, in this context, for the spectacularly inefficient pursuit of highly unpalatable small animals); but it was certainly the former, not the latter, who suffered the effects of the famine. The general point of this example is simply that it is aspects of the cultural phenotype that determine not only behaviour, but also susceptibility to extra-cultural forces.

The next logical step in this exposition is to point out that there are many forces impinging on humans (political, ideological, and, perhaps above all, economic) that are wholly cultural in origin. The impact of these forces, I suggest, is almost entirely mediated by the cultural rather than the biological phenotype of the individual. However, before developing this point, I want to consider a line of objection to my argument so far implicit in Cosmides and Tooby (1987).

Levels of Explanation

The general drift of my argument has been to claim that specifics of the cultural environment in which an individual is situated will provide determinants of behaviour that will tend to overwhelm any tendency towards biological optimality. But, as Cosmides and Tooby (1987) argue, the level of manifest behaviour is not the appropriate one upon which to conceive of natural selection as operating. It is rather that selection will optimize the structure and characteristics of the information-processing systems that mediate between environmental cues and behavioural responses. Thus they deny that there is any genuine conflict between cultural explanations of behaviour and an evolutionary theory of adaptive function. 'The claim that a behavior is the product of culture ... entails nothing more than the claim that surrounding or preceding individuals are an environmental factor that has influenced the behavior under discussion in some way. It leaves the learning mechanisms that allow humans to acquire and generate culture completely unspecified' (p. 293).

This is surely quite unobjectionable. In particular, it would be foolish to deny, and is probably important to assert, that the physiological mechanisms that enable us to lead the behaviourally complex, culturally varied lives that we do are a product of the evolutionary process. But despite their emphasis on the evolution of learning and information-processing mechanisms, as opposed to behaviours, Cosmides and Tooby do want to insist on the importance of this evolutionary perspective to the understanding of behaviour. The question that this raises, then, is to what extent the specifics of behaviour are best explained by appeal to the details of such cognitive mechanisms. I take it that Cosmides and Tooby do think that these mechanisms are highly relevant to behaviour, and that this is a major part of the force of their insistence on the domain specificity of cognitive mechanisms.

Cosmides and Tooby are clearly right to emphasize the necessity of an intervening level between evolution and behaviour. Evolution can only act on the physiological mechanisms that are proximate causes of behaviour. And they are therefore right to insist that the evolutionary approach to behaviour does not entail either uniformity of behaviour or uniformly adaptive behaviour. The conception of the mind as an all-purpose genetic fitness calculator is rightly derided, but its absurdity is no criticism of the evolutionary approach to behaviour. However, the problem that Cosmides and Tooby's approach raises, but does not, I think, fully answer, is to what extent the more complex picture they present really does substantiate a systematic connection between evolution and behaviour.

I shall, at least for the sake of argument, accept the following claims: First, that understanding the human mind requires functional analysis, and evolutionary theory is the most promising way of approaching such analysis. (The methodological issues here are admirably analysed by Kitcher 1987). No one could object to the use of evolutionary optimality analyses as a means of generating hypotheses about the functional organization of the mind; and with sufficient caution, the derivation from such analyses may provide additional weight in favour of such hypotheses.) And second, even less controversially, understanding the function and functioning of the human mind is an essential component of the explanation of human behaviour.

The reason that, in spite of these concessions, I remain sceptical about the relevance of evolution to human behaviour is that I do not believe, in general, that explanation is transitive. That is, from the fact that *A* is an essential ingredient in the explanation of *B*, and *B* in the explanation of *C*, it does not follow that *A* has a comparable importance in the explanation of *C*. And so, the explanatory relevance of evolution to cognitive mechanisms, and cognitive mechanisms to behaviour, does not imply that evolutionary theory has much explanatory relevance to behaviour.

Space will not permit an adequate treatment of this issue here (the position I have stated is defended in more detail in Dupré 1983[7]). An example should give some sense of the underlying issue. Levels of, and changes in, populations of various animals can be explained, to some extent, by models that reflect interactions with other predator and prey species (see e.g. May 1973). A standard case concerns the populations of lynx and hare in northern Canada. It is a presupposition of this model that lynxes are physiologically equipped to consume hares; and parameters in the model that measure their efficiency at doing so may be thought of as reflecting such things as the relative speeds of the two animals.

In this case, it is certainly true to say that these physiological properties of lynxes have an important influence on the prevalence of hares. However, in explaining the frequency of hares, such properties would not naturally be appealed to. The sort of property that one would rather want to know would be the propensity of lynxes to eat hares, presumably as a function of the frequency of hares (strictly a population-level property). That there is such a propensity at all certainly depends on lynxes having the appropriate physiological characteristics. But it is very questionable whether the required propensity could be inferred from physiological facts about the two animals; and even if it could, it is unclear what this would add to the explanatory project at the population level. I am inclined to put the point in the following, perhaps contentious, way: the abstract conception of a lynx appropriate for constructing population models has no tight theoretical connections with the conception derived from physiological studies of lynxes. The assumption of transitivity of explanations rests on the erroneous assumption that the models of an entity that are constructed for distinct theoretical purposes can be identified with one another. I suggest that they cannot.

Returning, then, to the specific topic under discussion, I do not take it as obvious that the model of the human mind derived from

---

[7] See also Dupré (1993, pt. 2).

an evolutionarily based functional analysis can be identified with that appropriate to the correlation of information absorbed from the environment with behaviour emitted by the organism. As in the ecological case just discussed, I suspect that the structure of the mind as conceived in the former project will provide important constraints on the possible accounts of the latter. Human physiology will limit what kinds of information can be absorbed, and what kinds of behaviour are possible for the organism. Evolutionary considerations may be highly relevant to the formation of plausible hypotheses about such matters. It might, for instance, turn out that people really do have very sophisticated capacities for distinguishing degrees of relatedness to themselves. If so, possibilities for patterns of social behaviour would exist that otherwise would not.

In sum, I agree with Cosmides and Tooby that evolution has structured the mind, but has not done so in a way that can be expressed in terms of behaviour. But I do not think this is simply because the mind determines behaviour in response to a variable environment. Rather, and this brings us back to the main theme of this essay, behaviour is a phenomenon that must be understood at the social level. The social cannot be constructed out of the psychological dispositions of individuals any more than the ecological can be constructed from the physiological. In both cases, the relevant properties at the more complex level of organization are, at least in part, constructed from phenomena at that level.

The Breakdown of Cultural Species in Modern Society

If the human species were divided into the relatively isolated tribal groups that are the classical subject matter of anthropology, I believe that the analogy I have been proposing between biological and cultural species would be almost exact. The adaptation of such groups to their environments is highly varied, and primarily varied with respect to culturally transmitted modes of behaviour. And such groups may well be as *culturally* isolated from their neighbours as, say, species of a typical genus of flowering plants are genetically.[8]

---

[8] Which is to say, not very. No doubt cultural traits have always been fairly readily transferable across cultures. The important point is not so much the extent of isolation as the extent to which particular groups, at particular times, exhibit distinctive and reasonably homogeneous forms of social organization.

However, in the modern world things are not so simple. To begin with, modern societies are stratified in ways that suggest that the taxonomy must be applied within, not just across, what are generally considered as cultures. For a paradigmatic class society, such as nineteenth-, and to some extent twentieth-, century Britain, such a classification may provide plausible cultural species; certainly such classes can be culturally distinct and very effectively isolated from one another.[9] But when we move to the contemporary United States it is doubtful whether such an analysis is useful. American society can be divided by economic class, ethnic background, religious belief, geographic region, and no doubt many other factors, all of which carry a number of behavioural propensities, and all of which more or less cross-classify the population. Perhaps a biological analogue could be constructed by finding a large area of land, temperate, fertile, and well irrigated, and introducing a few thousand fairly similar species of Compositae. I doubt whether, in ten years' time, a very useful taxonomy, or anyhow a unique taxonomy useful for all purposes, could be constructed for the consequences of such a scheme.

Since most of the human species lives in pluralistic rather than tribal societies, it may well be asked what the point of all this fuss about cultural species can be. But of course if contemporary culture is seen as historically resulting from the gradual hybridization of many earlier cultural species, it should be clear that this only emphasizes the importance of culturally transmitted properties. The point is rather that to whatever extent such primitive cultural species may have been natural kinds, there may be no such natural taxonomy for contemporary society.

How, then, does this illuminate the appropriate way of understanding contemporary human behaviour? Presumably, in the first place, we should recognize that the appropriate typology for modern humankind is going to vary according to the context of enquiry (it can be argued that the same is true for biological taxonomy generally, but that is not my present concern). And what this means is that what we need to consider are cultural properties, and the forces to which such properties make their bearers more or less susceptible. A suitable typology may of course be very helpful in understanding the distribution of such properties.

---

[9] Bourdieu (1984) provides a fascinating and detailed analysis of this kind for contemporary France. It appears from this example that a detailed cultural topography, at least, is feasible for some modern societies.

One example may suffice to illustrate the general point. A person's susceptibility to advertising might turn out to be a function of the amount of time spent watching television. One could then describe a pattern of consumption (behaviour, presumably) characteristic of the ideal advertising victim. This would approximately correspond to the amount of money spent advertising various products on television. It might turn out that the more time a person spent watching television, the closer their consumption approximated this pattern. It might even be possible to correlate the disposition to watch television with distinguishable cultural groups, perhaps religious affiliation. In this, obviously hypothetical, example, the disposition to watch television is an aspect of the cultural phenotype, and advertising is a cultural force the influence of which depends on that aspect of the cultural phenotype. It is also conceivable that there is a human typology that provides a good predictor of the relevant phenotypic feature. Whether or not this is probable in the present case, it is evident that such a taxonomy would be fairly specific to the particular problem in view. I take it that, hypothetical though it be, this is a fairly typical pattern of behavioural determination in complex modern societies.

A final complication is that in modern times there exist culturally generated forces that have intercultural and international dimensions. Consider, for instance, the recent precipitous halving of the price of oil.[10] This constitutes an enormous change in the flow of wealth between the producers and consumers of oil. Presumably it will eventually have profound effects on great numbers of people. Citizens of those relatively poor countries that depend heavily on income from oil production will on average become poorer, and consume less. Californians will probably drive their cars more, and Minnesotans keep their houses a few degrees warmer. The ramifications of such a phenomenon will at any rate be extremely complex, but in the end will be manifested in behavioural changes of numerous kinds. Cultural isolation, it would appear, is even less prevalent than might be imagined.

In the end, then, I suspect that the significance of the concept of cultural species is more historical than contemporary. But historical

---

[10] There are still, I suppose, some who would like to include such a phenomenon with biological evolution in a more generalized doctrine of optimality. That this particular change reflects a change in some ideal market, in any but the narrowest sense, would be a difficult case to make. That the independent decisions of buyers and sellers should be coordinated so as to clear the market may be an admirable thing, but is hardly a sufficient achievement for constituting Utopia.

significance is hardly to be demeaned when the underlying topic is evolution. If cultural speciation has the conceptual role in human evolutionary history that I am suggesting, then it should be no surprise that cultural properties and forces should now be the major determinants of human behaviour.

The Significance of Culture

I hardly intend to say everything that could be said about the topic at the head of this section. However, by way of conclusion, I would like to say something about why the emphasis on cultural forces and properties over biological analogues makes some very significant differences to the understanding of human behaviour. In particular, supposed political implications have figured very prominently in the debate over sociobiology, but I do not think that the issues have always been correctly conceived. I shall say very briefly here how I do think some of these issues should be treated. But before making some rather general comments on this topic, I want to address a more specific worry that might be raised about my present approach to this topic.

What I have in mind is the possibility that the concept of cultural species that I have advocated may be thought open to all kinds of disastrous political abuse. To assert fundamental differences between different human groups, it may be said, may provide ammunition for racist propaganda. Now of course, as I have emphasized, cultural species are not biological species, so no support is offered for the traditional form of racism that claims fundamental biological differences between human geographical variants. Indeed, to the contrary, it should be clear that cultural species will often cut across racial divisions. In the contemporary United States there are many, though perhaps deplorably few, people of African or Asian origin who are fully assimilated into contemporary American culture, or one of its predominantly Caucasian variants.

On the other hand, sophisticated xenophobes are inclined to talk more about cultural differences than about biological differences. Contemporary anti-immigration rhetoric, in particular, tends to appeal to fears of the local culture being swamped or overwhelmed by alien introductions. I do not believe, however, that the way to respond to such claims is to deny, in the face of the facts, the existence of cultural differences. Part of the response, on the other hand, should be to emphasize the actual nature of cultural diversity. Such diversity, particularly within pluralistic societies, will only be amenable to a very loose and probabilistic typography. The topology of such a typography will be a good deal flatter than that of typical biological variation. Certainly an essentialist conception of cultural groups would be even more glaringly inappropriate than it is for biological species.

But one crucial response to xenophobia is to make clear exactly what threat is supposed to be involved in the interaction of different cultures. This is the kind of consideration for which relatively simple models of cultural evolution may have considerable significance. For example, with respect to the question of immigration, it is difficult to see why the introduction of fairly low levels of unfamiliar cultural traits should be likely to spread through the population unless they prove to be both very attractive to the indigenous population and also consistent with the central themes and values of the existing culture. For instance, the introduction of a significant population of Indians and Pakistanis to the British Isles has no doubt led to a great increase in the consumption of the cuisines of these parts of Asia— a great boon to a gastronomically impoverished culture. I take it that only the most rabid cultural conservative would consider this kind of innovation objectionable. On the other hand, as far as I know there has been little tendency for the natives to adopt the Hindu or Muslim religions. (I do not, of course, wish to imply that these would be any less desirable than those presently dominant.) If cultural variants spread solely to the extent that they prove both attractive to the indigenous population and compatible with their antecedent cultural values, then such innovations would appear, prima facie, to be an unqualified benefit.

The most interesting possibilities perhaps arise from the consideration of density-dependent phenomena. It is certainly plausible that there will be a tendency for a culture to spread purely as a consequence of the numerical superiority, or power and wealth, of its adherents. I take it that in the United States, to whatever extent there has been assimilation of members of cultural and ethnic minorities, it greatly exceeds any reverse assimilation into the minority cultures. In so far as it is felt that cultural diversity is a good thing, certainly a reasonable claim, the real danger is in the excessive cultural assimilation of minorities—and perhaps, on an international scale, deliberate cultural imperialism—and not in the subversion of a culture by immigrants and minorities.

Turning now to the more traditional debate about the political significance of evolutionary arguments, much of the trouble derives from confusion over notions of determinism. Sociobiologists are widely accused of being biological determinists, or even genetic determinists, and respond with understandable puzzlement that they recognize many other forces, including environmental conditions, that interact with the alleged biological determinants of behaviour, apart from the purely biological. Moreover, the objection to determinism of whatever kind has to do with worries about the lack of human autonomy. It is often not at all clear how postulating cultural rather than biological forces has any effect on this problem.

However, the real problem here has not to do with any wholly implausible suggestion that all human behaviour is determined by some simple and homogeneous collection of forces. The point is rather the extent to which forces of particular kinds are susceptible to any kind of deliberate manipulation. And in this respect there is a great divide between cultural and biological forces. To whatever extent biological causes, genetically based, play a role in determining behaviour, they are susceptible to manipulation only in the very long term by behavioural eugenics. (I do not mean to deny, again, the dependence on the actual expression of such forces on other contextual factors. That is a quite different point.) Cultural causes, on the other hand, are, in principle at least, much more malleable. Lamarckianism, lateral transmission (within a generation), and asymmetric transmission are features of cultural evolution that suggest that, under favourable circumstances, change may be rapid.

But more even than just the pace of change, cultural evolution suggests no obstacle to the idea that change may be directed, and may even be directed in the service of objectives such as freedom and justice —or, of course, as may be historically more typical, oppression and injustice. (There is a philosophical line of thought, deriving from Kant, that asserts that this possibility is the only intelligible ground for human autonomy.) It is sometimes implied that such notions are only products of our particular cultural traditions or even, occasionally, of our biological evolutionary history. But, though there is something in this, since many of the most pressing, or depressing, contemporary examples of injustice have to do precisely with the way accidents of cultural history are taken as adequate grounds for unequal treatment, there are surely aspects of justice that can, and indeed must, be seen in a genuinely transcultural way. The deep truth that lies behind the political suspicions of sociobiology is that a biological explanation of the human condition may seem to explain what is wrong with the world, but does so in a way that suggests that this wrong is inevitable. Cultural explanation holds out the possibility not only of explaining, but also of changing the world. I do not mean to suggest that cultural forces are any less determinative of their good or bad causal upshots. It is rather that if we understood sufficiently how they operated, it would seem that they would be the kind of forces that should be susceptible to interactions with, and redirections by, deliberate acts of public policy. Of course, this is not an argument that such explanations are correct. That was the aim of the earlier sections of this paper.

The preceding remarks lead to a concluding thought on the role of the concept of optimality. Evolutionists (and some economists), despite increasingly numerous reservations, are still inclined to treat optimality as providing a methodology for describing the world. But given the way the world strikes most reflective people, there is much to be said for preferring to see optimality as concerned rather with a normative description of how the world should be. Although I do not mean to imply that economists or sociobiologists who defend the importance of optimality as a tool for descriptive analysis of human society typically confuse this with a normative conception of optimality, there is nevertheless a serious danger that the descriptive concept will seem much more significant—since more causally efficacious—than the normative. Emphasis on cultural determinants of behaviour, regardless of the extent to which they are usefully analysable by appeal to a descriptive concept of optimality, has the great virtue of showing how a normative conception of society might also have importance, and even causal weight, in our reflections on society.

7

Darwin and Human Nature

This paper is not about Darwin but about recent Darwinism. Though Darwin himself certainly had some ideas about my topic, human nature, here I shall be concerned not with these but with contemporary ideas deriving from Darwin, and the relevance of these for our understanding of human nature. I shall begin with some more recent history of science.

The evolutionary basis of human nature, or anyhow human behaviour, is a currently topical issue. Although, as is well known, attempts to apply Darwinian ideas to the better understanding of our own species have been made since soon after the publication of Darwin's theory, there has been a particularly active movement of this sort in the last three decades. The landmark event for this recent movement is, no doubt, the publication of E. O. Wilson's *Sociobiology: The New Synthesis* in 1975. Richard Dawkins's *The Selfish Gene*, published a year later, though in reality of rather limited relevance to human social behaviour, was probably much more widely read, and created a general if spurious impression that great advances in our understanding of evolutionary biology were making possible much deeper understanding of social behaviour, including human social behaviour.

A well-orchestrated splash accompanied the publication of Wilson's book, and since it was clearly designed more for the adornment of coffee-tables than for reading, few people seem to have been aware that only one chapter out of twenty-seven actually claimed much relevance to human nature. For those who really did want to read about Wilson's views of human nature, three years later he published a more obviously popular work entitled *On Human Nature*, which went on to receive a Pulitzer Prize. Though these books, and some other less notorious works at about the same time, staked out the territory for modern evolutionary studies of human behaviour, the academic reaction at this point was on the whole not very favourable. And by

This paper was first presented at the conference Darwin's Millennium, at the University of Southampton, July 1998. I thank the the organizers, Lucy Hartley and Cora Kaplan, for inviting me to participate, and the audience for stimulating responses.

the 1980s there were beginning to appear some scathing critiques of Wilson's project. Biologists such as Richard Lewontin and Stephen Jay Gould pointed out that most of human sociobiology wasn't really empirical science at all.[1] More often than not, sociobiologists would set out to explain some alleged human trait simply by making up a more or less plausible Just So story about human prehistory. Not only was there no reason to believe the stories, but often it turned out that there was little reason beyond stereotyped and clichéd assumptions to believe in the existence of the traits that were being explained. This was particularly the case for efforts to explain human sexual difference, always a central part of the sociobiological agenda. It began to seem that the Wilsonian human sociobiology was no more scientific than the previous generation of this genre represented by such works as Desmond Morris's *The Naked Ape* (1967); it was simply a lot less entertaining. By the mid-1980s 'sociobiology' was not an altogether respectable intellectual label.

One might very naively suppose that scientific projects exposed to devastating criticisms would just go away. Perhaps they sometimes do, but I fear the story here is a more typical one. Faced with the fact that 'sociobiology' was no longer a respectable label, sociobiologists did something much more practical: they changed the label. There are now hardly any sociobiologists. They have become evolutionary psychologists.[2] Again naively, one might suppose that they would have felt an obligation to rebut the criticisms to which they had been exposed. Instead, however, they have made the much smarter move of becoming truly scientific, and ignoring them.

Sociobiology of the Wilson generation was clearly presented and received as polemical. Unsurprisingly, therefore, it was subjected to counter-polemics. Perhaps because of its inherent deficiencies, it did rather poorly in these exchanges. Whether consciously learned or not, the lesson for the next generation of sociobiologists, or evolutionary psychologists, is obvious: don't engage in a polemical activity, just do the sociobiology. And this is what I mean by becoming scientific. No one expects chemists to spend their time debating with critics as to

[1] Some of the earliest critical responses can be found in Caplan (1978), including the notorious attack by the Sociobiology Study Group of Science for the People, whose members included Gould and Lewontin. The most detailed and incisive critique was provided a few years later by the philosopher of science Philip Kitcher (1985).

[2] Evolutionary psychology began to emerge from the wreckage of sociobiology in the mid-1980s. A definitive statement of its agenda and some of its claimed achievements can be found in Barkow *et al.* (1992).

whether the periodic table is well founded, or whether the structure of matter is really atomic.[3] They get on with their chemical research. And so with the evolutionary psychologists. Although evolutionary psychology does still have its professional apologists and debaters—sometimes philosophers rather than scientists[4]—it also increasingly has professional societies, technical journals, research centres, and so on. And of course if the typical evolutionary psychologist is a busy professional scientist he, or not infrequently she, will not want to waste time arguing with ignorant outsiders. One is tempted to say that evolutionary psychology has become, in Thomas Kuhn's famous phrase (Kuhn 1970), normal science. My only reservation about using this expression is that Kuhn clearly thought of normal science as generally a good thing: normal science, for Kuhn, really *is* too busy solving problems to spend much time debating fundamentals. From another perspective, evolutionary psychology looks more like what Kuhn called pre-paradigm science: it is competing with many other claimants to provide us with the fundamental approach to human nature and human behaviour. Here I am inclined to modify Kuhn. Kuhn thought that professionalization—journals, jargon, professional societies, etc.—were criteria for normal science. The present case suggests to me that the paraphernalia of professionalization are now, at any rate, quite compatible with a science that is, epistemologically, at least, pre-paradigm. Put another way, it seems that normal science can be faked as well as practised.

Of course, evolutionary psychologists do have a story that differentiates them from bad old sociobiologists. Specifically, they suggest that the trouble with the older, Wilsonian, sociobiology was that it was missing a crucial link between stories of selective advantage and contemporary behaviour, namely psychology.[5] They observed that selective regimes in the distant past could directly shape not behaviour, but only the brain. Thus they argued that Darwinian theory implies that the human brain should contain mechanisms apt to produce the

---

[3] The contrast I have in mind was nicely illustrated in a debate some years ago in the Senate of Stanford University as to whether the programme in feminist studies should be given degree-granting status. A chemist complained loudly that the programme appeared to take for granted a point the discussion of which, he felt, should be central to the agenda of such a degree, namely the question whether women were in any way disadvantaged. Background assumptions taken as too obvious for debate, evidently, are the prerogative of properly established sciences.

[4] A notable example is the philosopher Daniel Dennett, whose book *Darwin's Dangerous Idea* (1995) includes some uncompromising enthusiasm for evolutionary psychology.

[5] A good source for this argument is Cosmides and Tooby (1987).

kinds of behaviour that would have been selectively advantageous to our distant ancestors.[6] I must confess to being a little sceptical about the radical nature of the insight that separates contemporary evolutionary psychology from the bad old sociobiology. It seems unlikely that Wilson, say, imagined that evolutionarily selected behaviour was generated by some organ other than the brain. Perhaps there is a slightly sharper awareness of the possibility that useful behaviour in the Stone Age may be maladaptive now, though certainly that is not a wholly novel insight and again one that can easily be detected in Wilson's sociobiology. The emphasis on the mechanisms by which behaviour is produced, together with the trivial point that behaviour is produced in response to features of the environment, do certainly make clear that a mechanism that produced adaptive behaviour in response to Stone Age environments may produce maladaptive behaviour in contemporary environments. At any rate, this is the picture that forms the core of contemporary Darwinian accounts of human nature. This core is nicely summarized on the cover of a highly influential recent anthology edited by some eminent evolutionary psychologists, where we read:

Human nature can finally be defined precisely as the set of universal, species-typical information-processing programs that operate beneath the surface of expressed cultural variability; this collection of cognitive programs evolved in the Pleistocene to solve that adaptive problems regularly faced by our hunter-gatherer ancestors such as mate-selection, language-acquisition, cooperation, and sexual infidelity.[7]

It would be easy to spend many pages discussing these few phrases. Has it really been established that we are information-processing computers, as the 'precise definition' of our nature as consisting of a set of information-processing programs suggests? It is fascinating also to reflect on the curious choice of representative problems. Certainly only an evolutionist would think of sexual infidelity as one of the major

---

[6] There is also a question: which ancestors? We have an awful lot, and as sociobiologists like to remark, all of them were reproductively successful. The consensus favourites for this role are from the Stone Age or Pleistocene. This choice satisfies an obvious prerequisite of being a period after the split of our lineage from our closest relatives the chimpanzees. It is also alleged to be a period during which the conditions of existence were fairly static, and from reading sociobiology one has the impression that those conditions are now fairly well known. Both of these assumptions strike me as highly questionable, and this is a serious source for scepticism about the deliverances of contemporary sociobiology, though they will not be my focus in this paper.

[7] The paperback edition of Barkow *et al.* (1992).

problems of Stone Age life. But I shall leave sociobiology for a bit, and consider rather more generally the question of human nature.

Human Nature and Essentialism

One of the greatest conceptual implications of Darwin's theory is that it put an end to essentialism in biology. This has not always been easy to appreciate. Biology provides our clearest paradigm of a classificatory science, and scientific classification has been conceived since antiquity as involving the discrimination of kinds of things with distinct essences. A conception of science that is influential to this day is that its business is the discovery of the essences of things of different kinds and the elaboration of the processes by which these essences explain the gross familiar behaviour of things. The essence of a thing of a certain kind is conceived as both the property that explains the characteristic behaviour of a thing and the property that provides a necessary and sufficient condition for being a thing of that kind. Such a defining essence is conceived as timelessly determining the nature and membership of a kind. Nowadays it is often supposed that such an essence for biological kinds might be found in the genetic code, what might be called the bar code theory of genetics: feed a DNA sample into the sequencer and the machine will print out *Canis domesticus* or *Homo sapiens*. But what has become increasingly clear to post-Darwinian biologists is that there can be no necessary and sufficient condition for being an organism of a certain species, and that the characteristic properties of members of a species are, first, almost always typical rather than universal in the species and, second, to be explained in various different ways rather than by appeal to any simple or homogeneous underlying property. All this follows from the most basic Darwinian principle: the most fundamental characteristic of a species is its variability. It is only the fact that species encompass great diversity that makes evolution, and hence the very existence of species, possible. And genetic properties are just as variable as any others.

I mention all this because the expression 'human nature' looks suspiciously like a way of referring to the essence of the human species, and if that is right then there is surely no such thing. But perhaps this worry is a bit pedantic. Even if it is a mistake to suppose that tigers, say, have to be striped, that this is part of or follows from their essence, we can still say confidently that most of them are. If we think of the nature of a species merely as what is common or typical for members of that species, we will still have a useful concept, and one that will enable us to make more or less reliable predictions about organisms. The point about variability does have some important consequences, however. First, it is important to know that on no acceptable understanding of human nature does it follow from the fact that some feature is part of human nature that if something is human it must have that feature. This point is of particular importance in view of the fact that the natural is often given normative force. Homosexuality, for instance, has often been said to be 'against nature'. On a properly post-Darwinian understanding of nature, however, this can mean no more than 'non-typical'—a judgement that hardly justifies a condemnation. Second, and perhaps more important, these Darwinian qualifications of the idea of a species nature should make it clear that natures can change over time. What was once a rare property of members of a species may, over time, become common. It is sometimes supposed that the slow rate of evolutionary change makes this point largely irrelevant to time-spans of any practical interest. But this I think is a mistake. To explain why I must first discuss some much deeper and more pervasive contemporary misunderstandings of biology.

Contemporary Darwinism

The greatest difference between Darwin's own theory and contemporary Darwinism is the rise of genetics. Nowadays when we think of evolution we think of a synthesis, very roughly speaking, between the ideas of Darwin and of Mendel. Mendel's conception of discrete and heritable genetic units, by showing how features of organisms favoured by natural selection could be spread through and accumulated within a species, was necessary to convert Darwinism into a genuinely adequate account of the origin of species. At the present time, it is ideas descended from Mendel rather than Darwin that occupy more of the public mind—not to mention the public purse. Almost everyone has heard something of the Human Genome Project, for example, the multi-billion pound project to describe the entire chemical structure of human genes. However, there is a serious gap in this synthesis, and it is a gap out of which profound and important misunderstandings grow. This gap is the limited extent of our current understanding of development, or ontogeny.

Biologists think of genes in two quite different contexts. First they are substances that are passed on from parents to offspring, and the frequencies of which change as some parents leave more offspring than others. This is the evolutionary perspective on genes. But there is also a developmental perspective: genes are also thought of as guiding the process by which a fertilized egg becomes a fully formed organism. Much discussion both popular and professional in evolutionary biology brings these two perspectives together by speaking both of the genes for all kinds of traits, and of their increase in frequency under natural selection. But the intelligibility of all this talk is seriously compromised by the fact that we have such a limited understanding of the relation between genes on the one hand and the features of organisms that provide adaptive advantages on the other.

The underlying idea is, of course, that genes cause phenotypic features. But further analysis of this idea proves extremely difficult. Sometimes when we think of a cause we think of a sufficient condition for some event, as for example when we say that the impact of the cricket ball caused the window to break. Sometimes we mean a necessary condition, as when we say that exposure to a virus causes measles. In the first case the cause was not necessary, as the window might have been broken equally well by a baseball; in the second it is not sufficient, as many people exposed to the measles virus don't get sick. Philosophers have often flirted with the idea of something like a complete cause, which would be a set of conditions both necessary and sufficient for the production of an event, but it is very doubtful whether such a cause could ever be adequately specified. At any rate we can say with confidence that when biologists say that a certain gene is a gene for blue eyes, six fingers, aggressiveness, schizophrenia, or intelligence, they seldom suppose that the presence of that gene is a condition necessary or sufficient, still less both, for the trait in question.

A more recent philosophical idea has been the idea that a cause might be no more than something that makes its effect more likely.[8] A paradigm here is the relation between smoking and lung cancer. Most smokers, of course, never get lung cancer, and some non-smokers are unlucky enough to contract the disease. On the other

---

[8] I discuss this conception in more detail in Dupré (1993, pt. 3), and also provide further references.

hand what is abundantly clear is that the probability of acquiring lung cancer is enormously higher if you smoke than if you don't. This provides a good prima facie ground for the claim that smoking causes lung cancer. I say prima facie because there are great technical complications. The biggest problem, naturally exploited on occasion by sophisticated spokespersons for the tobacco companies, is that there may be something else, perhaps even a gene, that causes both the disposition to smoke and the disposition to get lung cancer. If this were so, the correlation between smoking and lung cancer would not point to causation, as indicated by the fact that deciding to smoke or not to smoke actually wouldn't have any effect on your likelihood of getting lung cancer.

This idea seems in some ways very well suited to the needs of genetics. Evolutionists are inclined to say that what 'gene for *x*' means is precisely that an organism with that gene is more likely to have the trait *x* than is an organism without it. Since organisms do not simply have holes in their chromosomes, an organism without the gene has some other gene (or allele, that is, an alternative version of a gene) in its place. Thus this definition applies only to genes of which there are several alleles available. It also applies only to features which are not universally present in the species, since nothing can increase a probability of 1. This definition is adequate for the purposes of evolutionary theory, since only where there is both genetic and phenotypic variation is there any question of evolutionary change through natural selection. But once a gene or a feature becomes for the time being fixed in a species, the definition ceases to apply. So the idea of causation in question is largely irrelevant to the understanding of development.

It is not difficult to see what is wanted. Think again of the lung cancer case. Perhaps what we have in mind when we say that smoking causes lung cancer, and what is ruled out when we discover that smoking and lung cancer are joint effects of a common cause, is that there is some mechanism that ends in lung cancer and in which smoking plays an important or essential part. This, at any rate, is the sort of claim we want to make for the genetic case. Such mechanisms will be very complex and involve many factors other than the gene with which we are concerned. Even for a relatively simple trait like eye colour, the production of brown pigment, say, in the eye will require the cooperation of several genes and the presence of a very large number of other chemicals in the appropriate cells.

Now consider what is involved in the claim that a certain gene is a gene for intelligence. (And for the sake of argument let me suppose here that there is some measurable quantity to which 'intelligence' refers.) There are, of course, countless genes that have the potential to make a positive or negative effect on intelligence. Genes that build brains properly will generally tend to produce more intelligent organisms than inferior or defective alternatives. And no one doubts that very large numbers of genes have important roles in building brains. But no one doubts either that countless environmental factors also play a role in the determination of human intelligence or, for that matter, in the determination of the fine structure of the brain. Richer, more stimulating environments early in life, parental inputs of many kinds, better teachers in smaller classes, and so on. Can we at least assume that genes that tend to increase intelligence will, if intelligence is an adaptive advantage, tend to become commoner? Not at all. To start with, not only do many genes cooperate in the production of a phenotypic trait, but it is well known that single genes often play a role in the production of many traits. Such other traits may have disadvantages that outweigh the advantage provided by greater intelligence. Second, we must be cautious about the very idea of a gene that increases intelligence, *tout court*. It is often the case that a gene that has one effect in one context has an opposite effect in another. So really we should speak only of genes for intelligence in a particular environmental, and also genetic, context. And third, selective advantages are always relative to a context. In the present case we might speculate, for instance, that if the environment were to become static and predictable this would increasingly favour mechanical instinctive responses over intelligence.[9] Thus both the selective advantages of intelligence and the tendency of a gene to promote intelligence are transitory conditions that can change rapidly.

This is all well known, even banal. What is remarkable, however, is the extent to which reflection on these elementary points undermines an enormous amount of what is often taken for granted about genetics. Perhaps few people now believe in the kind of genetic

---

[9] It is very interesting that sociobiologists believe that the human tendencies they study arose during a long static period of prehistory, an assumption that lends some credibility to the idea that evolved mental tendencies should be relatively inflexible in their expression. This historical assumption has been called in question lately; if it should be rejected, we would have some reason instead to predict the evolution of more flexible dispositions. For an interesting discussion of the dependence of the advantages of intelligence on the vagaries of the environment, see Sober (1994, ch. 3).

determinism that holds behaviour to be somehow or other directly produced by genes. But in the slightly weaker sense of holding that our behavioural dispositions, and perhaps our ultimate goals, are genetically determined, genetic determinism remains widely influential. This influence is evident in the metaphors so widely used for talking about the genetic. Think for example of the phrase 'genetic code'. Originally, I suppose, the term 'code' was used to refer, perhaps harmlessly enough, to the relation between triplets of DNA bases and amino acids. This is not, of course, literally a code, for the relation is causal not semantic. Nevertheless, it is easy enough to see the appeal of the metaphor. Nowadays, however, it is not uncommon to hear, at least among the less biologically sophisticated, the idea that genes code for eyes or noses. This, of course, is just nonsense.

The more sophisticated are liable to related confusions. A metaphor that is of great interest here is that of genetic information. I call this a metaphor because of the connotations, many of them misleading, that this term carries. There are, on the other hand, technical meanings of 'information' that are correctly and literally applied to DNA. The central technical meaning of information has to do with the reduction of uncertainty. A paradigm of a very large and measurable amount of information is the winning number of the national lottery. Knowledge of this number reduces the possibilities for which ticket will win the lottery from millions to one or a handful. Because information theory works most straightforwardly with specific numbers of initially equiprobable alternatives, it is easy to see why it has become associated with the digital computing revolution. A digital computer is an enormously complex network of binary options any state of which is, prior to all programming, equiprobable. As we move into domains with less well-defined sets of alternatives, the application of the concept becomes more problematic. Still, the general idea that the causal influence of genes can be seen as reducing the uncertainty of developmental outcomes continues to have merit. There are, however, problems that derive from the metaphorical connotations of the term 'information'. Most obviously, information has come to be associated in most peoples' minds with chunks of text. Training in information technology is intended not merely to enable people to have some causal influence on the world thereby reducing uncertainty, but more specifically to enable them to move structured sets of symbols about. But the suggestion that there is something symbolic about the genetic code is entirely spurious.

A more subtle point is the following. The idea of genetic information is frequently used to support a view that might be referred to as 'genetic exceptionalism'. By this I mean the idea that DNA has a role in development different in kind from that of the other causal conditions with which it interacts. Biologists sometimes express this idea in the terminology of information theory by suggesting that DNA carries information for which the environmental conditions are merely a channel. But within information theory this is simply a mistake. In this technical sense very many chemicals in the cell, the grosser physiology of the mother's body, and, particularly in the case of humans, all manner of subsequent interpersonal inputs all carry information for which the DNA may be considered part of the channel. That is to say, simply, that all such factors reduce uncertainty about the developmental outcome. This unwarranted privileging of the DNA leads via a verbal play on the term 'information' to the idea that the DNA is something symbolic—a map, a blueprint, a coded message, etc. It is an interesting speculation that it might have been much more salutary for genetics if rather than drawing its metaphors from semantics, with the suggestion of a static relation between sign and thing signified, it had appealed rather to hermeneutics and thereby drawn attention to the constantly interactive or dialectical relation between genotypes and phenotypes.[10]

All this relates also to the prevalent notion that the genetic is the source of largely immutable outcomes. This flows naturally from thinking of the DNA as providing, for example, a blueprint. But one need only recall that the strongest defensible causal interpretation of the claim that there is a gene for some trait is that the probability of that trait is higher given the presence of the gene, to see that there is in general no inference from the presence of the gene to the inevitable expression of the trait. This does not dispose of all the problems common to contemporary claims that there are genes for rape, intelligence, homosexuality, obesity, male desertion of children, and so on, but it should at least moderate the impression sometimes conveyed that these

---

[10] The preceding two paragraphs are greatly indebted to the book *The Ontogeny of Information*, by Susan Oyama (1985). The implications of this important work are only slowly becoming clear to those interested in evolutionary theory. Peter Godfrey-Smith (2000) convincingly argues that there is a legitimate understanding of DNA as embodying a code in relation to its role in the economy of the cell, but that this metaphor cannot properly be extended beyond the role of genes in the production of specific proteins. As I understand it, this argument does not affect the substance of what I am claiming here.

highly developmentally complex characteristics should be seen as flowing inexorably from secret messages in the language of DNA.

Let me now get back to evolution. Much of what I have said is relatively uncontroversial among careful analysts of biological ideas. It is certainly not controversial that DNA interacts in highly complex ways with a variety of cellular, physiological, ecological, and social factors in the course of ontogeny, and it is not very controversial that this simple observation immediately disposes of many of the more hyperbolic claims made for the possible growth of genetic knowledge. A little more controversial is the claim that a realistic consideration of the relation between genotypes and phenotypes makes implausible the perspective on evolution popularized for a while by Richard Dawkins, which sees all natural selection as the selection of genes. There is a theoretical perspective from which one might see this genic selectionism as a possible bookkeeping device, as a way of tracking the genetic course of evolution, though even this is of very questionable practical relevance to any real biological study. But given that science is interested in causation, it seems clear that this perspective is inadequate. It is not entirely clear how much the Dawkinsian perspective has to do with the issues of present concern. However, it is certainly the case that many popular and even not so popular presentations claim Dawkins's insights as a central plank in the genetic determinist case, so it is worth digressing to indicate some of the difficulties with this perspective.

The most serious objection to Dawkins's perspective on natural selection is that, once again, it fails to give an adequate account of the complexity and diversity of causal relations in evolution. The classic illustration of this point has been familiar for almost forty years, and concerns a gene referred to as the t-allele in the house mouse.[11] This gene has a number of interesting properties. Firstly, it is very bad for individuals: in individuals homozygous for the gene (that is, with two copies of it) it is either lethal or causes male sterility. Though this obviously causes selection against individuals possessing the gene, counteracting this is an unusual property of the gene known as meiotic drive. Contrary to familiar Mendelian principles, male mice with only one copy of the t-allele nevertheless produce sperm almost exclusively containing the t-allele. These two processes allow the prediction of an expected frequency of the t-allele at selective equilibrium. And here is the most interesting part of the story: the observed frequency is

---

[11] The original research is due to Lewontin and Dunn (1960). A good discussion is in Sober (1984b: 262–3).

significantly lower than that predicted. But an explanation for this fact is quite widely accepted. Mice live in small groups known as demes. Sometimes it will come about that all the males in a deme will be homozygous for the t-allele, and therefore infertile. In this case the females in the group will be effectively sterile, not because of any individual properties they possess, but simply by virtue of belonging to the wrong group. This is a paradigm case of group selection. We thus have three distinct causal processes that determine the frequency of this gene. Selection against individuals and against groups is counteracted by meiotic drive, the last a true case of gene selfishness. A bookkeeping measure of the fitness of the gene fails to capture any of the causal complexity of such a case. It is now quite widely acknowledged that selection acts at many causal levels.[12]

Despite the increasing realization of the complexity and diversity of biological processes, orthodox biologists still hold to a version of the synthesis of Darwin and Mendel that reached its definitive form at mid-century. One idea that holds this orthodoxy together is the central position of some relatively simple models of population genetics, models that describe the changes in frequency of genes under various evolutionary regimes. I am sceptical of the usefulness of such models and their relevance to real biological systems.[13] I am also sympathetic to some recent moves that break much more radically with the Darwinian tradition. A number of biologists, for instance, have expressed scepticism about the sufficiency of natural selection to account for the evolution of highly organized systems, and have explored the possibility of integrating Darwinian ideas with recent thinking about self-organization in complex systems.[14] In a rather different direction, taking a step beyond the kind of pluralistic selectionism I have just discussed, is a move further away from the primacy of the genetic under the rubric of developmental systems theory.[15] This has generated a rapidly growing body of discussion in the philosophy of biology. The central idea is that we should go the whole way with decentring the gene and consider impartially the whole set of developmental resources involved in reproduction broadly construed. Dawkins's emphasis on the genetic leads him to see a chicken as fundamentally an egg's way of making another egg. Developmental

---

[12] See Sober (1984b), and for a detailed treatment of group selection, Sober and Wilson (1998).
[13] This scepticism is elaborated in Dupré (1993: 131–42).
[14] See esp. Kauffman (1993) or, for a more accessible exposition, Kauffman (1995).
[15] This is the perspective deriving from Oyama (1985). For a good survey, see Griffiths and Gray (1994).

systems theorists, while not denying this, insist that the egg is equally the chicken's way of making another chicken, and a bird is even, just as legitimately, a nest's way of making another nest. Life consists, on this view, of numerous and diverse cycles of development and reproduction, involving an often very large variety of different resources. As far as I know, no serious attempt has yet been made to apply this idea to human evolution. But in principle it is clear that such an application would involve not only talk of genes and life in the Pleistocene, but resources that have much more recently become central to human development such as schools and hospitals. It is, of course, unlikely that such a development would be amenable to the kind of mathematical modelling so much beloved of a certain scientific caste of mind, and represented in evolutionary theory by the models of population genetics. But perhaps that mathematical incalcitrance represents merely the complexity of the subject matter.

I have emphasized that much has happened in evolutionary theory since Darwin wrote his famous book. But of course this should not be taken to imply a negative evaluation of Darwin himself. Darwin's status in biology is perhaps unique in the sciences. Almost any theoretical claim about biology routinely appeals to Darwin as a precursor, and where there is a serious theoretical debate, Darwin, like Nature, is enlisted on both sides. Contrary to this tradition, some of my comments here have suggested sympathy for major movements away from Darwinian orthodoxy—though at the end of this chapter I shall return to biological orthodoxy by attempting to enlist Darwin in my support. As a matter of fact, it seems to me entirely likely that in a hundred years' time we will see Darwin as fundamentally wrong about most things, and natural selection, while surely an important biological process, as much less central to biology than it is seen today. This does not, of course, cast aspersions on Darwin's contribution, any more than Newton's scientific reputation should be jeopardized by the achievements of Einstein. My strongest reason for suspecting that Darwin will not, in the end, be seen to have got it right is that that, I think, is how science works. Perhaps ultimately science will converge on the truth—though this is a controversial claim. But if it does so, it will not be through the sudden insight of an innovative thinker such as Darwin, but from centuries of increasingly dreary graft. I seriously doubt whether great scientists ever get things right: if there comes a time when it is possible to get things right, it will be a time that no longer needs great scientists.

Human Nature Again

The reason I have spent so much of this chapter on these general issues about the current state of evolutionary biology and genetics is that this is what one must come to terms with in order to make a just evaluation of the achievements and also the excesses of contemporary Darwinism. Much of evolutionary psychology, in particular, seems to me a strikingly dogmatic application of what are in truth a set of developing, partially controversial, ideas. Indeed evolutionary psychology is often presented in a tone that goes beyond the dogmatic to verge on the evangelical. And this seems to me inappropriate for any part of science, but even more so for a science which is still in many respects at an early stage in its theoretical development. (A deeper understanding of this dogmatism might require a thorough exploration of the thorny matter of the equivocal relation between evolution and religion.)

What does follow from Darwinism for our understanding of human nature? I have already noted that on certain understandings of human nature what follows is that there is no such thing. But let us mean by human nature something that is typical of rather than universal in our species, and something that may be typical of it only at a particular historical moment. I don't want to deny that there are some aspects of humans answering to this characterization. Our minds are not blank slates: we are social animals, and almost all of us acquire a remarkable facility with language; probably no form of socialization will make many of us consent willingly to a life of slavery or celibacy. No doubt more could be said. But these are, I take it, banalities, drawn from our common knowledge of the human species rather than any sophisticated grasp of biology.

The obvious objection to drawing any morals about human nature from biology is that human nature does seem to change rapidly through historical time and across cultural space. This would suggest that not much about human nature follows merely from our individual biology. Now it is, as I have noted, part of our biology to be social, so that to say that social factors play a large role in determining the particularities of human nature at a time is not to put the human species outside biology. A correct understanding of this point is easily derived from the perspective mentioned above of developmental systems theory. Being highly—indeed in important respects uniquely —social creatures it is not surprising that the developmental cycles by which we are reproduced contain massive inputs of a social character. Presumably it is the malleability of this contribution to human-nature-at-a-time that accounts for the speed at which human nature changes over historical time.[16] And of course, given this conception, it is a very important possibility that there may be intentional input into these changes such that human nature may perhaps, in the long run, be a product of human creation.

A common response to such a picture is to insist that even if behaviour is produced by an interaction between what we are and the environment in which we live, still there is a common core of what we are. And it is this latter that a scientific account of human nature should be primarily concerned with. To see what is at stake here, we must return to more abstract biological ideas. A concept that has been very important in shifting the hold of genetic determinism on biological thought is that of a norm of reaction. The norm of reaction is the set of responses of a particular genotype to a range of environments. A standard example is the reaction of a plant to various soil and weather conditions. The size of a plant of a particular genotype will vary with these environmental factors. And different genotypes may do better in different environments. And so on. Though this has been a very important concept, it is typically developed in a way that retains a certain kind of gene-centredness: it is just that the genes facilitate a defined range of behaviours rather than a single mode of behaviour. And in this respect developmental systems theory represents a more radical rejection of this gene-centredness. Part of what is at stake with gene-centredness is a reductionist perspective on science. The overwhelming focus on the gene would not, I think, seem so plausible unless we were already in the grip of the picture that saw the individual as primary in the explanation of human behaviour: the individual is the source of behaviour and the internal structure of the individual is the way to understand the behavioural dispositions of the individual.

I don't propose here to take on this perspective in a general way. Rather, I want to urge a way of thinking about human nature and human behaviour that is radically at odds with the perspective in question. Darwin, in *The Descent of Man*, emphasized the deep sense in which humans were a social species. I suggested above that human nature might be more profoundly social than that of any other kind

---

[16] This line of argument against the atavism so characteristic of evolutionary psychology is developed in more detail in Dupré (2001).

of animal. Let me put on one side the question of the social insects, with only the remark that it is open to doubt whether a bee or an ant should be considered an individual at all, rather than a small part of a much larger organism, the hive. At any rate the comparisons I have in mind are with social animals that are, beyond doubt, also properly seen as individual organisms.

Recall the idea I want to oppose: individual organisms have a set of capacities and dispositions fully determined by their internal nature, perhaps ultimately their genetic nature. Which capacities are actually exercised will depend on triggering conditions provided by the environment. Against this, I will argue that the capacities possessed by a human (and perhaps to some degree by other social animals) are not intrinsically determined, but are themselves constructed in large part by the social context.[17]

An illuminating way of developing this point is to reflect on our concept of disability. In line with the internalist perspective that I am opposing, it is natural to think of human disability in terms of intrinsic incapacities: the inability voluntarily to move one's legs, to discriminate parts of the visual field, and so on. But as philosophers have long emphasized, action is not the mere movement of one's limbs, but the intentional accomplishment of some task in the world. We move much closer to the latter when we think not, for instance, of the ability to move one's limbs, but rather of the ability to move around one's environment. But without denying that voluntary control over one's limbs is, for many, a significant aid to moving through the environment, it is neither a necessary nor a sufficient condition for possessing the locally typical degree of mobility. In an environment fully equipped with lifts and ramps, a person with no control over their legs but a good wheelchair is as able as anyone else to move about. In the vicinity of Los Angeles, it is far more important to this end to have possession of a car than it is to have the use of one's limbs. Thus what are ultimately social decisions (construing 'social' very broadly, as I don't mean to discuss the means by which societies actually make decisions) determine what kinds of people have what kinds of capacities. The intrinsic conception of disability is, of course, important from a medical perspective. Medicine is rightly concerned with providing for people as large as possible a range of the intrinsic capacities characteristic of our species. Nevertheless, the relation between intrinsic capacities and actual capacities to act is mediated by social conditions.

[17] This argument is developed in more detail in Dupré (1998).

And it is capacities of the latter kind that are more fundamental to our understanding of human action.

The preceding points have important implications for contemporary neo-Darwinian accounts of human nature. Evolutionary psychologists attempt to define human nature in terms of behavioural dispositions evolved by our ancestors in the Pleistocene. These dispositions are quite specifically adjusted to particular domains of behaviour that are taken to have been of exceptional evolutionary significance, constituting what evolutionary psychologists like to refer to as psychological modules. It is assumed that we know enough about the Pleistocene to make compelling arguments about what would have been advantageous to our ancestors during that period.

It must first be stressed, contrary to an impression one can still receive not only from popular evolutionary writings, that successfully demonstrating that a particular property or disposition would have been useful to an organism is very far from implying that organisms of that kind must have evolved that property. Almost equally difficult is the problem of providing evolutionary explanations for observed traits.[18] Even given the presence of a trait, the hypothesis of the selective scenario which might have given rise to it is fraught with difficulty. But for the case of human behavioural dispositions the problem is far worse, because it is a matter of great difficulty to establish that such dispositions exist at all. Consider a notoriously controversial case, the alleged evolution of a male disposition towards rape.[19] It is reasonably plausible that among our distant male ancestors a genetically based disposition to rape females might have been selected. Reasonably plausible—but quite impossible to establish with any confidence; still less is it possible to establish that such a genetically based tendency actually existed to be selected. If it could somehow be established that a deep tendency to rape existed almost universally among contemporary human males, it would lend some support—though still qualified—to the evolutionary speculations. But most human males do not commit rape, and the allegation that they are nevertheless innately disposed to do so rests on nothing but the presentation of a more or less plausible evolutionary scenario. We have a small and epistemologically worthless evidential circle. Attempts

---

[18] The classic source for these difficulties is Gould and Lewontin (1979).

[19] This hypothesis is defended by, among others, Thornhill and Thornhill (1983, 1992). For criticism, see Fausto-Sterling (1985) and Ch. 9 in this volume.

to provide empirical evidence for this disposition, as for example by means of the wonderfully named 'objective phallometry', in which men's penises (most often the penises of prison inmates) were attached to measuring devices while the subjects watched rape movies, offer parodies of scientific method rather than evidence.

But here I want to advocate what is perhaps a deeper point. It is that the whole project of theorizing extra-cultural human behavioural dispositions is ill conceived. This is, I suppose, more obvious in the case I have just been discussing than in the examples of disabilities I discussed earlier, though for somewhat different reasons. The possibility of committing rape, where by rape we mean something that happens in contemporary human societies, is not something that can even exist independently of any social context. This is why sociobiological claims to have observed rape among ducks or flies tend to strike people who have thought seriously about the human problem of rape as ludicrous. No doubt it is possible that our male Stone Age ancestors copulated with their female contemporaries against their will. But this is hardly the same thing as an accusation of rape against one of our contemporaries. Rape currently involves such things as: a framework of law and its violation; conceptions of the rights of individuals; historical relations to conceptions of the ownership of women by men, and more recent attempts by women to insist on their own ownership of their bodies; and so on. The complex social-constructedness of rape is clearly demonstrated by current debates about the extension of the concept of rape to marital rape or date rape. And it is little more absurd to say of a duck merely that it committed rape than it is to say that it is guilty of date rape. Rape, like most kinds of human action, is partially a social construct. (It should perhaps be added that this, like countless other possible examples, shows how strange is the common view that to be socially constructed is somehow not to be real.) The point can finally be related back to developmental systems theory. The kinds of cycles whereby forms of human behaviour are defined and reproduced are very different in many ways from those of biological evolution. Some of these differences have been explored formally by theorists of cultural evolution. For now I note only that once one escapes from the narrow view of human ontogeny embraced by certain reductionist perspectives on human biology, it is easy to see that many forms of human behaviour are conditioned by processes that differ fundamentally from those of natural selection.

Evolutionary psychology, then, I take to be deeply flawed both epistemologically and conceptually. If this is indeed the best attempt that can be made at connecting human nature with evolution, it proves to be a very disappointing project. Where, to conclude, does this leave Darwin's relevance for our understanding of human nature?

With perhaps an important exception for some consequences of sexual selection, the idea that the evolutionary origin of humankind should illuminate human nature does not figure prominently in Darwin's work. Certainly he was much concerned to demonstrate the common origins of humans and other animals, and no doubt this alone has broad if not very specific implications for human nature. It was important in the nineteenth century and is still important in some places today that there is little reason, once the basic premiss of evolution has been accepted, to believe that a soul or immaterial mind is a likely possession—or unique possession—of the human species. And more generally, the commitment to some sort of continuity between humans and other animals provides an important heuristic for assessing explanations of human behaviour.

Few would suggest that Darwin's primary importance was as an inspiration to late twentieth-century evolutionary psychologists even if some encouraging remarks might be found by dredging through his corpus. However, in keeping with the unhistorical tone of this chapter, and also with the practice mentioned above of invoking Darwin's support in any theoretical discussion of a biological issue, I shall suggest that whatever Darwin's actual views on the subject, the implications of his work are on the whole quite antagonistic to current evolutionary psychological aspirations.

Perhaps the most important reason for this has already been mentioned, the fatal consequences of Darwin's views for essentialism. Though it would be an exaggeration to call most contemporary sociobiologists essentialists, their project does often fail to take sufficient account of the fact that the demise of essentialism makes it quite unclear whether there are many historically stable species-wide traits of the kind they hypothesize. One way to look at the situation is to contrast the real timelessness traditionally associated with essentialism with the effective timelessness of the traits proposed by sociobiology. In evolutionary time a few million years is of course the blink of an eye. But in terms of human plans and aspirations it is rather long; and a process of change that requires the timescale of genetic evolution is, apart from explicit engagement in eugenics, effectively unachievable.

A central part of my present thesis on this subject is that we have no reason to make any such assumption, and the complexity of the determinants of human nature that I have been trying to indicate are such as to leave quite open the possibility that there might be processes of change on a much briefer, and more practically relevant, timescale.

But ultimately I want to enlist Darwin on my side of this debate on more general grounds still. Prior to Darwin were religious world-views and scientific–mechanistic world-views. These two fitted well together through the familiar image of God the mechanic who reached his ultimate statement in the work of writers such as William Paley. It is a familiar idea, still reverberating today, that Darwin under-cut in the most profound way to date the theological world-view. It is a less familiar idea that he also provided the death blow to the mechanistic world-view. Mechanism sees everything that happens as, ultimately, implicit in the arrangement of matter and the causal laws governing the changes that matter undergoes. Evolution, on the other hand, is a theory of *transcendence*: objects come into being with properties that could not have been envisaged simply from a knowledge of matter. It is, on the other hand, a *naturalistic* theory of transcendence. No immaterial or spooky substances or properties are called upon to explain this transcendence; nevertheless, the con-sequent is more than the sum of its antecedents.

This is a controversial characterization of evolution, though it is one I am happy to defend. What I want to suggest further is that just as biology naturalistically transcends its material basis, so human nature naturalistically transcends its biological base. The ground of this transcendence, obviously enough, is culture. Sociobiologists consider such appeals to culture as metaphysical mystery-mongering. My suggestion is that they see things this way because, like many other scientists, they remain trapped in the reductionistic, deterministic metaphysics of Darwin's scientific predecessors. A true appreciation of Darwin's philosophical contribution, not to mention a more sophisticated view of biology that that philosophical insight helps to make possible, shows the inadequacy of this mechanistic scientific approach to human nature.

V
Gendered People

8

Sex, Gender, and Essence

Introduction

The primary aim of this paper is to show why the assumption that real essences underlie the kinds we distinguish in scientific investigation is mistaken. I want to claim that this assumption is not merely empirically unwarranted, but necessarily at odds with a genuinely empirical approach to science. Briefly, unless one supposes the discovery of a kind to imply the discovery of an essence, there is nothing more to the discovery of a kind than the discovery of the correlations of properties characteristic of the members of the kind. Since I do not believe that essences are to be found so easily, I shall argue that the importance of the discovery of kinds to the progress of science is much less than is generally supposed.

Although I shall illustrate this argument with a fairly detailed discussion of some specific classificatory concepts, I would like to emphasize at the outset that the role of this discussion is just that— illustration. It would be possible to understand the basic thrust of this argument by reading only the introduction and the conclusion of the paper. However, the cases that I shall discuss should certainly assist in the understanding of the argument (as well, I hope, as having some intrinsic interest of their own). The significant point about the concept of sex is that although it is undoubtedly a concept that has major significance for biology, and although it is also a concept that divides the natural world into well-defined classes, the scope of the generalizations to which it gives rise is at every stage an empirical matter. In its briefest outline, the force of my argument will be that if a real essence is to serve any purpose, it must at least determine the scope of generalizations covering the entities that realize it. But, for a serious empiricist, there is never any reason to suppose that this can be done.

I would like to thank the Pew Memorial Trust for a grant that supported the initial stages of this research, and the Stanford Humanities Center, where it was completed.

The case of sex is particularly revealing because despite the extreme sharpness, by biological standards, anyway, of the cleavages it makes in the organic world, it turns out that there is neither evidence nor reason to expect that it gives rise to any generalizations across the broad categories that it defines. The concept of gender, a concept that has received a good deal of elaboration and clarity from the last fifteen to twenty years of feminist scholarship, has been developed in ways that demonstrate that one cannot assume that the intersection of biological sexual categories with some smaller biological category will give one the appropriate scope for generalization. Even as applied to one species, *Homo sapiens*, the scope of generalizations restricted to male or female is a purely empirical matter and, in most interesting cases, far narrower than that of the entire species.

I shall begin the paper with a brief explanation of what I understand by essentialism, or, at any rate, of the sense in which I shall claim that it is objectionable. I shall also mention some immediate difficulties that it presents in the context of biology. I shall, then, discuss first sexual categories, and then gender categories, as illustration of the impossibility of extrapolating from the existence of a kind to the scope of any generalizations about its members. In the discussion of the latter categories, I shall mention some of the reasons for confidence that the sex–gender distinction cannot be dissolved. In conclusion, I shall argue that the cases considered point to fundamental defects in the essentialist point of view, and that this point of view should be abandoned rather than modified.

Essences

In thinking about essences, it is first necessary to distinguish two very different functions that they may be supposed to serve. First, essences are often conceived as properties that determine the answer to the question to what kind the object that instantiates them belongs. But, second, essences are also thought of as determining the properties and behaviour of objects that instantiate them.

Within the first type of function, we can introduce the familiar Lockean distinction between real and nominal essences. The view that nominal essences determine the kinds to which objects belong amounts to little more than the introduction of a bit of technical terminology. A nominal essence is connected to a kind by some sort of linguistic convention. Since it is obvious that we could not refer generically to members of any kind without the existence of some linguistic convention determining the (at least approximate) limits of that kind, the existence of nominal essences as characterized above is not controversial. More detailed description of nominal essences may certainly commit one to more or less powerful semantic theses —for example, to the view that there must be necessary and sufficient conditions for the application of every classificatory term—but will carry no metaphysical commitments.[1] My concern in this paper will be solely with real essences.

To assert that there are real essences is, in part, to claim that there are fundamental properties that determine the existence of kinds that instantiate them. The existence of such properties will have profound metaphysical consequences: in particular, it will imply that the existence of kinds of things is as much a matter of fact about the world as is the existence of particular things. Such kinds are quite independent of our attempts to distinguish them, and their discovery is part of the agenda of science. It is consistent with at least the majority of modern usage to take the previous sentence as providing a necessary and sufficient condition for the existence of a *natural* kind, and I shall hereafter use that expression in that way. It is important to note, however, that although the existence of a real essence is, then, sufficient to determine the existence of a natural kind (ignoring possible problems about non-instantiation), it does not follow that a real essence is necessary. As a matter of fact, I believe that there are natural kinds without real essences, unless perhaps in an almost vacuously attenuated sense.[2]

I would also like now to emphasize a point that will be central to the argument I have in mind against essentialism. This is the observation that even if a kind *is* determined by a real essence, it is hard to see what route there could be to the *discovery* of essences other than the prior discovery of kinds. This should immediately lead one to entertain serious doubts about the empirical credentials of such essences. In particular, if this epistemological point is correct, we should

---

[1] This claim is, of course, controversial in the light of the well-known views to the contrary of Kripke (1972*b*) and Putnam (1975*c*). I have argued against these views elsewhere (Ch. 1 in this volume). The possibility of deriving essentialism from semantic considerations has also been attacked at length by Salmon (1981).

[2] This is a reasonable way of interpreting the conclusion I defended about species in Ch. 1.

be very suspicious of any practical consequences that appear to follow from the existence of a real essence. In the remainder of this section I shall develop the concept of a real essence so as to try to show that if the existence of a real essence amounts to anything, it does, indeed, have practical consequences—specifically, in entitling us to anticipate the existence of laws governing the behaviour of objects that partake of it. In the following two sections I shall illustrate the fact that the differentiation of kinds entitles us to no such anticipation. The conclusion that will be developed in the final section of the paper is that discovering kinds does not involve discovering essences; and so, given that there is no other way of discovering them, nothing does.

However, to return to the main thread of the argument, it does seem that, barring the most radical and implausible nominalism, there must be *something* to the doctrine of real essences as so far described. Some kinds are, at the very least, more natural than others. The class of creatures with wings and feathers, for example, is more natural than that of creatures that are grey and over one foot long. This is so because when we know that a creature belongs to the first class, we can make numerous further reliable predictions about it—that it, or its female relatives, lays eggs, is warm-blooded, and so on. Membership of the second class carries no such benefits. Depending on how deeply we can explain such clustering of features, we can adduce more or less powerful characterizations of real essences. If, to take one extreme case, God had simply chosen to assemble creatures in the light of some preconceived ideas of which features went well together, the real essences might amount to no more than conjunctive, or perhaps partially disjunctive, descriptions of God's aesthetic preferences. Since these descriptions would still reflect genuine clusterings of properties, they would at least be natural kinds and would exhibit, in a sufficiently weak sense, real essences. However, we naturally believe that the discontinuities in nature admit of somewhat deeper explanations, and this leads us finally to the second, and much more problematic, function of real essences, that of explaining the nature of the members of the kinds that such essences determine.

The strongest possible notion of a real essence would be that of a property, or group of properties, that determined—and, hence, in principle could be used to explain—all the other properties and behaviours of the objects possessing them. Although such a notion *might* be defensible for individual essences (Locke seems sometimes to have envisaged the microstructural description of an object potentially playing such a role), it cannot work for the type essences that are my present concern. This is for the simple reason that there is no kind (with the possible exception of the ultimate microphysical kinds) the members of which are identical with respect to all of their properties, even their intrinsic ones. I say 'intrinsic' properties because it is obvious that the *behaviour* of an object will typically depend on both its intrinsic properties and its external environment. Clearly, the strong view I am considering should claim only that essence determines behaviour as a function of some set of external variables, or, in other words, determines precisely specifiable dispositions to behaviour. However, variation in intrinsic properties requires a more fundamental retreat from the strong position. Specifically, some distinction between essential and accidental properties, that is, between properties that can and those that cannot vary between members of a kind, is unavoidable.

A promising and natural modification of the strongest conception of real essence, which provides a way of drawing just the distinction mentioned above, is the following: the essence of a kind determines just those properties and dispositions of its instances for which it is a matter of natural law that members of the kind will exhibit those properties or dispositions. The essential properties of members of a kind will then be, first, the real essence itself, and, second, those properties and dispositions nomically determined by the real essence. The rest will be accidental. Thus, for example, it is clearly no law of nature that squirrels are grey, since many are black. On the other hand, perhaps it is a law that squirrels have tails, and, hence, tailedness is an essential property of squirrels. An essentialist holding the position I am now suggesting would explain this by saying that the essence of squirreldom, perhaps a particular genetic structure, determined the growth of tails, but not a particular colour of coat. The suggestion that the essence might be the genetic structure illustrates another important aspect of such a position, the way that the essence itself is to be distinguished from other essential properties. Presumably, the genetic structure causally determines the growth of a tail, and not vice versa. Thus, the essence itself is that property that is explanatorily primary among the set of essential properties.

Making such an essentialism applicable to any part of biology—and, probably, to most other parts of science—requires some important qualifications. To begin with, it is very difficult to find really sharp distinctions anywhere in biology; generally, there is a range of intermediate cases. Certainly, as far as taxonomic distinctions are concerned, sharp boundaries are the exception rather than the rule. Thus, a theory of essences would have to be considered as applying to typical members of kinds rather than to all members. Assuming that it remains desirable to attribute individuals to kinds despite their abnormality, the laws applying to such kinds could be only probabilistic. The probability that something has a tail, given that it is a squirrel, would then reflect the frequency of the abnormality of taillessness. The modified essentialist position could be maintained by insisting that there is, nevertheless, some standard genetic structure that constitutes the essence of squirreldom, and that anything that perfectly realizes this structure would, as a matter of nomic necessity, have a tail. Less-than-ideal squirrels would then be judged to be squirrels, or not, on the degree of similarity of their genetic structure to this standard form. It will be apparent, however, that this unavoidable modification leaves the essential–accidental distinction rather more arbitrary than might have been wished. It is, at least, unclear whence the fundamental difference between blackness and tail-lessness of squirrels derives—apart from a patently question-begging appeal to the essential nature. If it comes to no more than a quantitative difference in frequency, then a fairly arbitrary decision is required to include one (or its genetic basis), but not the other.

Many philosophers of biology would wish to mitigate this difficulty by denying that taxonomic distinctions define kinds at all, and a fortiori, that they are natural kinds that could admit of real essences.[3] I myself believe that species (higher taxa I assume to be purely nominal kinds) are kinds rather than individuals. However, I doubt that the present problem will be much mitigated by denying that species are kinds. Whatever kinds one happens to favour in biology (e.g. ecological or evolutionary), one is unlikely to find the sharp boundaries that would evade the present difficulty. A striking illustration, which will be discussed in detail in the next section, is that of sex. It would be hard to imagine a more obviously natural division within biology than that between male and female. Yet in sexually dimorphic species

---

[3] Classic statements of the view that species should be treated as individuals have been made by Hull (1976) and Ghiselin (1974). An excellent sense of the present state of the debate can be gleaned from Kitcher's (1984) attack on the view and Sober's (1984a) reply.

there are typically variations with respect to sexually specific characteristics, and even genuinely intermediate individuals. At any rate, the distinction between fairly sharp boundaries between kinds and absolutely sharp ones is itself an absolutely sharp one, so that the advocate of biological kinds that completely evade the present problem will have a difficult task.

Part of my reason for emphasizing this difficulty is to stress the detachability of a belief in natural kinds from a belief in essences. The belief that there are discontinuities in nature to be discovered rather than invented is quite independent of the question whether these discontinuities are sharp or gradual. Moreover, the relation of natural kinds to questions of explanation does not depend on a doctrine of essences. One might suppose, for example, that there was some optimal set of laws (perhaps maximally deterministic and/or explanatory) governing a domain, and that the classes of entities recognized by those laws should be considered as natural kinds. Such a view does not require that any fundamental distinction be drawn between the essential and accidental properties of the members of such kinds. Since it will typically be the case that the frequencies of such properties in a kind will vary continuously from almost 100 per cent to almost 0 per cent, such a distinction appears inevitably arbitrary. But, as I have tried to show, without this distinction, the point of essentialism becomes obscure.

I shall return to the idea that natural kinds should be treated strictly as derivative from the discovery of laws in the final section of this paper. But, for now, I shall move on to discuss more specific cases in detail. This will demonstrate some further and compounding difficulties with the essentialist perspective.

Sex

As promising an essential distinction as one is likely to find in biology is that between male and female. The distinction can be drawn successfully for a very large number of organisms, and although, as I have suggested is true of almost any biological distinction, there are borderline cases, the vast majority of organisms of types to which the distinction applies can be assigned unambiguously to one category or the other.

Another relevant feature of the distinction, although now only for as long as we look at a particular type of organism, is that there are systematic differences between males and females at various levels of structural organization, and that these are causally and explanatorily related. More specifically, for most species, males and females differ genetically, physiologically, and behaviourally; and we are fairly confident that the genetic differences cause the physiological ones, and that the physiological differences cause the behavioural ones.

However, further consideration shows that the situation diverges greatly in certain respects from the essentialist scenario I sketched in the previous section. The properties that are causally fundamental in explaining sexual dimorphism between the members of a species are unquestionably not the properties that realize the real essences (if any) of maleness and femaleness. A microstructurally oriented essentialist might be inclined rashly to assume that the essence of maleness and femaleness for humans was/is the possession, respectively, of an XY or an XX chromosome. But many animals that can be divided into males and females as clearly as humans can have no XX or XY chromosomes. Indeed, this view would seem to imply that to say that there are both female humans and female geese would be a gross equivocation on the word 'female', since in each case the word refers to a quite different microstructural property; and this would patently be absurd.

Surely the correct way of describing the situation is to say that *for humans*, having XY or XX chromosomes *causes* individuals to be male or female. What it is to be male or female, on the other hand, is a property at a higher level of structural organization, that of producing relatively large, or small, gametes. It is *this* distinction, based on the fact that most types of organisms have individuals of two kinds distinguishable by a major dimorphism in the size of the gametes they produce, that is referred to by the general categories of male and female, and that in particular species is caused by a particular genetic dimorphism.[4] Thus surprising, and even paradoxical, though it may seem, it is correct to say that physiological differences between the

---

[4] An interesting paper by Michael Lavin (unpublished), primarily addressed to some philosophical problems that arise from gender reassignment surgery, includes a persuasive argument for the view that what we mean in ordinary language by 'male' and 'female' has nothing to do with either genetic or general biological considerations, but is derived wholly from considerations of gender, that is, of socially constructed conceptions of what it is to be male or female in our society. Although I am entirely sympathetic to this view, I hope it is clear that it is these more technical considerations that are relevant to my present discussion.

sexes, and any genetically determined behavioural differences that there may be, are not, in fact, caused by the sex of the organism; rather, these differences and the sex of the organism are joint effects of a common cause.

In this light, it is *not* surprising that the sexual categories have little explanatory power. It is very doubtful, that is, whether there are any very significant laws relating to males and females in general. It seems plausible that every generalization about a sexually specific characteristic is limited to some narrower group than that of all sexually dimorphic species. In some cases, there is a recognized taxon over which the generalization applies, either because the character concerned is an evolutionary novelty in a phylogenetically demarcated taxon, typically a species, or because that very character is used to define the higher-level grouping, as with mammals, or placental mammals.

Although the possibility cannot be ruled out a priori that there might be some properties universally, or almost universally, correlated with large or small gamete production, there seems to be no reason to expect that this will be the case. This observation invites reconsideration of my claim that sexual categories are exceptionally promising candidates for biological natural kinds. The intuitive basis for that claim certainly has nothing to do with a knowledge of laws pertaining to males and females in general. It is based, rather, on two kinds of observation. First, that within any species, and often within much larger taxa, there are very pervasive sex-specific generalizations to be made. Men grow or shave beards, and women have breasts; males and females of large numbers of (related) species have relatively similar genitalia. And, second, for enormous numbers of species, it is possible to distinguish males from females. However, what these observations properly suggest is that sex is a very significant property that may be appealed to in the analysis of innumerable different taxonomic groupings but that, nevertheless, it is not a property that is sufficient to define any significant kind. Alternatively, if one wishes to insist that males and females *do* form natural kinds, then there are natural kinds with little or no explanatory power.

The fact that nature can be 'carved at the joints' without yielding explanatorily significant categories is worth a moment's reflection. The explanation in this case is not hard to find, deriving from a very fundamental fact about biology: biological kinds reflect historical similarities as much as they indicate similarities of causal power. The divide between males and females, as general categories, derives not from characteristic properties or dispositions of the two classes but, presumably, from the existence of a very pervasive evolutionary tendency towards sexual dimorphism.[5] But it seems likely that the common evolutionary pressure may do no more than favour a simple dimorphism of gamete size, and that subsequent elaborations of the dimorphism may well be much more specific to particular evolutionary lineages, and not susceptible to large-scale generalization.

Two responses to the preceding argument need to be considered. First, I have so far ignored a trend in contemporary biology that *does* want to maintain the general explanatory power of sexual categories. By this, I mean a major area of sociobiology. And, second, one may accept the general conclusion that I have argued for above and yet explore the possibility of defining narrower, but still explanatorily powerful, sexually delimited kinds. I shall now briefly discuss these positions. The second will lead conveniently into the topic of gender.

I cannot hope to give an adequate treatment here of the highly problematic and controversial discipline of sociobiology.[6] However, one major area of sociobiological theorizing does assert precisely what I have said there is no reason to believe: that the simple fact of gamete size dimorphism strongly disposes species to certain subsequent evolutionary developments, specifically, to quite well-defined behavioural dimorphism. At its most general, the theory asserts that those organisms with smaller gametes (i.e. males) will tend to develop behavioural strategies that maximize the dispersion of their gametes, while the females will develop strategies that tend to increase the chances of successful development for those offspring that they are able to assist. At this very general level, the theory is based simply on the idea that a large gamete is a more significant investment of resources than a small one, and this will give disproportionate encouragement to strategies that tend to further its development. If there is any force to this argument, there is obviously a lot more to it when the reproductive physiology of the organism requires that much larger investments of resources are demanded for the female to have any chance of reproduction, as is the case of viviparous animals or animals that

---

[5] The nature of this pressure, however, remains surprisingly obscure. Excellent sources on the problem are Williams (1975) and Maynard-Smith (1978).

[6] The classic text on sociobiology is Wilson (1975); a highly readable popular introduction is Dawkins (1976). The enterprise has come under devastating attack from Lewontin *et al.* (1984) and, perhaps a little more sympathetically, from Kitcher (1985).

lay large eggs. Additionally, in such cases it is argued that when the offspring, or egg, is produced a substantial time after fertilization and requires further care to have any chance of survival, the female will find herself playing an evolutionary game with no cards. The male, it is argued, will by then have taken off to attempt to impregnate more females, and the only way that a female can expect to have any reproductive success at all will be to provide at least the essential minimum of parental care.

The most obvious defect with this argument is that the predictions to which it gives rise do not turn out to be true. Many species, even of birds and mammals, are quite monogamous in both sexes,[7] and there are many species in which the male provides as much parental care as the female, or even more. But I shall not attempt any evaluation of the general force of this sociobiological argument, since the preceding simple observation is sufficient to demonstrate the conclusion I wish to draw for my present purpose. This is simply that however significant a *force* in evolution these arguments may indicate, that is all they indicate. Clearly, if there is such a force, it is one capable of being overridden by other forces that operate in an opposing direction; otherwise, there could not exist the many exceptions just mentioned. (The same point can, and will, be made in connection with alleged systematic differences between men and women.) It might be thought that since I have allowed that dispositions common to members of a kind suffice to give that kind explanatory power, the above concession would be sufficient to constitute sexual categories as natural kinds. But this would be a confusion based on the failure to distinguish historical from causally explanatory categories. It may be that in every species there has been an evolutionary tendency for males to acquire dispositions to promiscuity and females to acquire dispositions to parental care. But in many cases those dispositions have not been actualized; and, hence, the members of many species do not have those dispositions. A drake, say, may have no disposition whatever to desert his mate. And it would be absurd to say that he must have such a disposition merely on the grounds that his ancestors had some, in fact unrealized, tendency to evolve such a disposition. So, in short, whatever the force

---

[7] Research over the last fifteen years has suggested that monogamy is a good deal rarer in both sexes than had previously been supposed. This in no way affects the general point about the great variety of sexual behaviour.

of these sociobiological arguments, though they may help to explain the particular behavioural dimorphisms in particular species, they do nothing to make males or females into genuinely explanatory kinds.[8]

The second response I described above was to accept that sexual categories are not themselves explanatory kinds but to argue that more narrowly defined sexually specific categories might, nevertheless, be so. Thus, male and female mammal, goose and gander, and man and woman may constitute sex-specific natural kinds with explanatory force regardless of whether male and female are themselves such kinds. Two general points should be made about this proposal. First, assuming, which I would be very reluctant to do for any taxonomic level above the species, that the taxon that is being sexually restricted is a genuine kind, this is not a case of the intersection of two kinds, but one of the subdivision of one kind. This is simply the application of the main conclusion of the present section about general sexual categories. But second, there is certainly no a priori objection to the internesting of natural kinds. There is nothing incoherent, for instance —though there is, almost certainly, something false—in conceiving biological taxonomy in this way. Metal and iron provide one plausible example. Human and woman might be another.

It is worth mentioning that the viability of this proposal will depend on accepting that some taxonomic groupings are, indeed, natural kinds. For one who believes that species are individuals (see note 3), it would be quite extraordinary to suppose that these individuals might be formed from the union of two kinds. However, this is not the place to pursue that issue.

It seems that there is nothing deeply wrong with this idea provided one registers some important qualifications. In particular, it would be absurd to suppose that man and woman, say, were 'better' kinds than human. To admit that species are kinds is to admit that kinds may encompass very considerable variation and, hence, license only probabilistic nomic generalization. Moreover, as I have argued earlier, it is to admit that kinds can be considered as defined by essences only in the most attenuated sense of 'essence'. It would, again, be absurd to suppose that the essence of woman was any more clearly definable than the essence of human. But, in fact, one would predict the opposite. Since the similarities between men and women are vastly more

---

[8] Kitcher (1985, esp. 166–76) shows clearly the internal weakness of this sociobiological argument.

numerous than the differences, one would expect the latter kinds to be 'worse'. And the problems with defining an essence of woman must surely, then, be more severe than those of defining an essence of human.

My final point follows, once again, from the fact that male and female are not themselves explanatory kinds. The explanatory significance of sexually specific kinds must be wholly empirically determined. No systematic differences between the males and females of a particular species can be assumed beyond those that are used to distinguish the sexes. This does need slight qualification. Sometimes one can appeal to higher-level generalizations. If one discovers a new species of mammal, one will reasonably anticipate that dissection will reveal an approximately familiar and sexually dimorphic type of reproductive physiology characteristic of mammals. However, I know of no other type of property for which, in the case of higher animals at least, such broader generalizations would be of any use. Certainly, there are none in overt morphology beyond similarities of external genitalia in related taxa; and more importantly, there are none in the area of behaviour, or again, none that extend beyond very narrowly defined phylogenetic groups. And, as I have insisted, none can be deduced from the mere fact of subsumption under the broad sexual categories. Accepting, then, the possibility, if highly qualified, that species as kinds may be subdivided into sexually specific subkinds, it is now time to look in more detail at the human case and to turn to the topic of gender.

Gender

The term 'gender', as it has been developed in contemporary feminist theory, refers to the sexually specific roles that are occupied by men and women in various societies. The most obvious reason for insisting on a sharp distinction between sex and gender[9] is that whereas

---

[9] Some feminists, notable among them Alison Jaggar (1983: 112), now want to resist drawing such a distinction between sex and gender, on the grounds that it erroneously suggests that the sexual side of the dichotomy is rigid and unchanging; and that, in fact, there is a continuous dialectical interaction between cultural and biological aspects of gender differentiation. Nancy Holmstrom (1982) develops a similar position and defends the conception of a distinctively female nature, on the basis that 'nature' should be understood in a way that encompasses both biological and culturally determined aspects, since she also denies that these can be intelligibly disentangled. Although I do not want to take issue with this view and willingly disavow any implication that there is some readily distinguishable set of immutable biological differences between men and women, I believe that my appeal to this distinction in the present context is both useful and harmless.

whatever properties may follow from the sex of their bearers, such as reproductive physiology and secondary sexual characteristics,[10] must be equally prevalent in all societies, it is quite clear that gender roles, on the contrary, are highly variable and culture-specific in many respects. So even if man and woman as biological categories are modestly explanatory natural kinds, it is clear that much of the behaviour encompassed under gender roles is no part of what they explain.

It is not altogether easy to assess the *extent* of variability in gender roles. One reason for this is that a great deal of the relevant research is particularly susceptible to the kinds of problems to which feminist critics of science have drawn attention; if there is any part of science for which the accusation of distortion by male-biased preconceptions seems particularly plausible, this is surely it.[11] But both anthropological and historical evidence leaves little doubt that such variability is extremely widespread.[12] Some particularly noteworthy areas are those that have been of special prominence in attempts to reduce gender-specific behaviour to a causal consequence of sex. Promiscuity, and the extent to which it is a male prerogative, provides one important example. Also of interest is the variability in the extent to which the generally socially approved form of gender-specific behaviour is adhered to or insisted upon. The prevalence of, and attitudes towards, homosexuality and incest, both subjects that have received a great deal of attention from sociobiologists, appear to be highly variable. The prima facie evidence seems to be that in most of the aspects of behaviour that suggest sexual dimorphism in the context of a particular culture, there is a great deal of cross-cultural variation.

Before continuing this discussion, it will be useful here to recapitulate a little, and explicitly to reintroduce the topic of essentialism. One traditional view might be the following. Both humans and males

---

[10] Secondary sexual characteristics—for example, the distribution of body hair—in fact show considerable geographic variability. If Darwin was right in attributing the majority of geographical variations among humans to a process of sexual selection (see Darwin 1981, esp. chs. 7, 19, 20), this is hardly surprising.

[11] See e.g. Longino and Doell (1983); Reed (1978).

[12] A good illustrative source is the collection of essays in Ortner and Whitehead (1981). It should, perhaps, be mentioned that these authors have more interesting and ambitious goals than merely establishing gender role variation. Ortner and Whitehead's introduction *begins* with the sentence 'It has long been recognized that "sex roles"—the differential participation of men and women in social, economic, political, and religious institutions—vary from culture to culture.' Nevertheless, for anyone who doubts this claim, these essays include ample evidence.

constitute natural kinds with a certain essential property. To be a male human is to partake of both the relevant essential properties, and much of the behaviour of a male human can be explained by reference to the causal powers of one or both of these essential properties. Against this I have argued that the most that can be sustained is the claim that male humans form a subkind of humankind. If this kind has an essential property, it is presumably a combination of the essential genetic structure of humans with the specifically sex-determining genetic features of male humans. The reason that the essentialist is forced into this specific, and, I suspect, rather unpromising, form of genetic determinism is precisely that maleness in general is not an explanatory category, and the only available candidate for an explanatory essence for the kind of human males must be their distinctive genetic features. Unpromising or not, there are certainly those, certain sociobiologists providing their theoretical wing, who want to maintain a position of this kind and trace the behavioural differences between men and women to the genetic.

A major thrust of the feminist research that has emphasized the historical and anthropological variability of sexual differences in behaviour has been explicitly directed against positions of this kind. Its aim has been to establish that these differences are to be understood in terms of social forces, which are fairly specific to particular cultures. It has also offered alternative schemes of explanation, perhaps the most influential and interesting of which are those that trace these differences primarily to the action of economic forces and conditions.

It would be an oversimplification to suppose that feminist scholarship fits uniformly into this agenda. To begin with, there are some feminists who would pretty much accept the essentialist structure that I have just outlined, while objecting only that the details have been filled out in a way revealing profound male bias. Most noteworthy in this category are a small number of feminist sociobiologists.[13] But more significantly, a markedly essentialist flavour has often been detected in a good deal of more mainstream feminist thought.[14] Indeed, it may even be suggested that the very intelligibility of feminism depends on construing women as a natural kind and, hence,

---

[13] See e.g. Hrdy (1981). At a more popular level, an entertaining feminist answer to Desmond Morris is Morgan (1972).

[14] Jaggar (1983) suggests that a commitment to biological determinism is a characteristic defect of the school of feminist thought she describes as 'Radical Feminism'.

on accepting essentialism. Although there may be some feminist projects that do, indeed, depend on this assumption, in most cases such suspicions are ill grounded. It will be worth a short digression to indicate why this is so.

It is easy to overestimate the prevalence of essentialist assumptions in feminist writing by failing to identify its primary goals. A great deal of emphasis in feminist work has been accorded to one observation that strongly appears to be a cross-cultural universal, namely, that men seem invariably to have achieved a position of domination over women. I do not think it would be unfair to say that this is often seen as the central theoretical problem of feminism. And if the central theoretical problem is one of explaining a universal fact about the relation of men and women, it is not surprising that much of the writing has a rather essentialist flavour.

But it is crucial not to overlook the fact that feminism, perhaps more than any other area of academic interest, is at least as much a political movement as a theoretical enquiry. From a political point of view, the universality of male domination is clearly of paramount importance. The political achievement of feminism may be described without exaggeration as the discovery and definition of an entire political class ignored by traditional theories. Nevertheless, the political significance of patriarchy should not blind us to the fact that from the point of view of the purely theoretical task of understanding sex and gender categories, this fact is anomalous rather than central. A brief consideration of the reasons for this will also help to forestall any tendency for the universality of male domination to serve as a motivation for a crudely biological theory of gender differences.[15]

Although I would readily concede that if male domination is a universal or near universal phenomenon throughout human societies, it is a phenomenon well worth theoretical study, there is no reason whatsoever for taking this as contradicting the basic variability in human sexually differentiated behaviour. In the first place, it is only one case to set against many. But even this way of putting the case is misleading; male domination is a phenomenon on a higher level of abstraction than is the characterization of particular forms of

---

[15] As that, e.g., of Steven Goldberg (1973). Unfortunately, it is also my impression that some feminists have been led by the same observation in the same direction, though certainly those who, like Goldberg, see male aggressiveness as the crucial, and even biologically grounded, factor are likely to point out that aggressiveness is not necessarily an unqualified virtue.

behaviour in particular societies. There is no reason to suppose that the exercise of male domination is itself something that has always been implemented by the very same kinds of behaviour. On the contrary, the kind of labour that women perform for men is quite different in, say, feudal societies, hunter-gatherer societies, and modern industrial societies; and the social institutions and personal interactions that enforce such performance are equally variable. Hence, the *implementation* of male domination—which is what we should consider, rather that its mere existence, if we are evaluating the plasticity of behaviour—far from contradicting the variability of gender roles, graphically illustrates it. Analogously to my conclusion for the case of sex, there may be good reason to suppose that human sex differences give rise to forces that have some tendency to bring about male supremacy. But, as in the previous case, although the existence of such a force may be of use in explaining the genesis of a particular gender-differentiated society, it does not pick out any property that characterizes the present state of that society. In this case, even if male supremacy is genuinely universal, the enormous variability in the form that it takes indicates extensive interactions with more specific forces that, in turn, show that there are no grounds for assuming that even the abstractly characterized consequence is in any way inevitable.

Returning now to the main theme of my argument, as in the broader case of sex in general, the question we should consider is whether even man and woman (in their biological sense) are genuinely useful explanatory categories. I have conceded that as purely biological kinds, they are largely unobjectionable; it may be allowed as a modest nomic generalization, for example, that humans born with penises will tend to grow facial hair later in their lives.[16] But there is a very powerful tendency to extend the relevance of explanatory categories beyond their empirically determined limits—a tendency, I am suggesting, that derives philosophical nourishment from the idea that when one has distinguished a kind, one has discovered an essence. If, in fact, the empirical significance of the kinds man and woman does not go beyond some systematic, if quite variable, physiological differences and the observation that men appear to have achieved a dominant position in all or most societies, the kinds distinguished seem of very modest significance. Certainly, nothing in

---

[16] For a fascinating account of the limitations of even such biological generalizations, and of the extent to which sexual dichotomy is socially enforced, see Fausto-Sterling (2000).

those empirical facts provides the slightest motivation for thinking that these categories should be accorded fundamental importance in explaining the particular forms of behaviour found in very different social systems, whether such explanation is motivated by sexist apologetics or (misplaced) feminist ardour.

The conclusion I want to defend might be stated as follows. Just as the concept of sex in general will do very little to explain why peacocks, but not turkeys, have long tails, or why the prairie chicken, but not the goose, is polygynous, so the notion of woman will do nothing to explain why oriental women once had their feet mutilated, or why twentieth-century Western women are more likely to become nurses than doctors. In principle, the same move is open to one as was suggested at the conclusion of the discussion of sex in the previous section. It would be possible to suggest that the appropriate explanatory categories were again to be narrowed, so that for behavioural explanations, the relevant classes would be as specific as female !Kung or male Spaniard. At this point, however, the claim to have identified even the most attenuated natural kinds would be impossible to sustain. The claim to have identified a natural kind must involve the idea that the behaviour of its instances depends, in some cases, on intrinsic properties of the individual characteristic of members of that kind. But it would be hard to find even the most bigoted racist nowadays prepared to assert that the social interactions characteristic of a man raised in, say, rural Spain would have been just the same if that individual had been brought up in a wealthy California suburb. To concede that explanations must appeal to kinds with that degree of specificity is to concede beyond serious argument that it is local, presumably cultural, factors that determine the relevant forms of behaviour.

Morals for Essentialism

I would like to take the discussion of the preceding two cases to illustrate a general argument against essentialism. The main thrust of this argument is a plea for complete empiricism with regard to the explanatory potential of particular kinds. My suggestion is that a belief in real essences either is vacuous or violates this demand. In partial reaction to this point, I shall also suggest that attention be drawn away from the attachment of fundamental importance to the delineation of kinds, and directed towards the identification of properties, dispositions, and forces. To connect these points, what makes a kind explanatorily useful is that its instances share the same properties or dispositions and are susceptible to the same forces. But since we have no way of deciding how much of such concomitance to expect in any particular kind, the discovery of a kind adds nothing to the discovery of any correlations that may turn out to characterize it. An essence, as I characterized it in the first section of this paper, can be seen as a promissory note on the existence of such correlations. It is a promissory note that empiricists should reject. I take the preceding discussion to illustrate this point in the following way: it is easy enough to distinguish classes at many different levels of generality—males, male vertebrates, men, Irishmen, and so on—but there is nothing in this process of differentiating classes that provides any basis for predicting the extent to which its members with be amenable to lawlike generalizations. Finally, this in no way impugns the theoretical significance of the *properties* on the basis of which such classes are differentiated (I shall elaborate on this remark with regard to sex below).

The most powerful example I have offered in support of this plea is the case of sex in general. As I said in the course of discussing sex, what we see is that there are major seams in nature that not only fail to distinguish robust natural kinds, but also fail to distinguish classes that realize any general lawlike regularities. The explanation in this case is simple enough: the seam reflects a presumably uniform type of historical process rather than the discrimination of any causally uniform type of entity. But it is hard to see what could be the basis for postulating the existence of a natural kind, in the strong sense of a set of common possessors of a real essence, except either the perception of natural seam among phenomena, or the discovery of one or more laws satisfied by a class of phenomena. In the first case, as the example of sex shows, the inference to a natural kind would be illegitimate; and in the second case, unless it constituted a quite ungrounded assumption that further hitherto undiscovered laws were in the offing, it would be wholly redundant.

Having rejected the idea that general sexual categories could provide a basis for lawlike generalization, I then considered the possibility that far more restricted sexually specific categories might still constitute natural kinds in the strong, essentialist sense. I do not mean to claim that this possibility has been—or, for that matter, could have been—rigorously explored in the space I have allowed. However, the issue of gender shows, at least, that it cannot generally be assumed that such kinds are discoverable. Here again, it is not difficult to see what is going on. Many factors affect human behaviour, including behaviour that is gender-differentiated. Looked at in this obvious way, it would be foolish to assume that there were forms of behaviour that were determined simply by the agent being, say, a human male—still less by the agent being merely a male. Nevertheless, unless we are careful to restrict the import of our categories to the empirical, we are in danger of being led into just such an assumption.

I should emphasize that I do not take myself to have shown any particular limits on the nomic significance of sexually defined categories in general. In the case of humans, there are good reasons for doubting whether this significance extends beyond the purely physiological. In many other species, there are undoubtedly good generalizations to be made about sexually dimorphic behaviour. My point is not that sex is a scientifically useless concept, but rather that from a conceptual standpoint that seeks kinds and their underlying essences, one is very likely to misrepresent that significance. I should now, therefore, say something very briefly about how I do understand its significance.

To begin with, nothing I have said contradicts the idea that sex is a highly significant *property*. By that, I mean that sex is a property that, in a sufficiently specific context, is frequently susceptible of lawlike generalization. At a certain time of year, for instance, the males of a particular species of bird produce very characteristic and predictable noises. If you know the time of year and the species of bird, what you additionally need to know, if you want to predict whether it will make that noise, is what sex it is.

No doubt of more theoretical interest are the ways that sex connects with evolution. At the most theoretical level, there is the question of the origin and maintenance of sexual reproduction. Since sex seems, prima facie, such an extraordinary waste of reproductive energy from the point of view of most females, this is a very baffling question. On the other hand, this very problematic nature of the phenomenon makes it likely that there is some powerful evolutionary process at work. At a less general level, as Darwin emphasized at great length, the existence of sex can have profound effects on the particular course of the evolution of a species. Thus, I am far from denying the biological interest of sex. What I want to claim is that the way in which the basic sexual categories—male, female, neuter, hermaphrodite—divide the natural world tells us nothing about either the extent to which such categories will give rise to general laws or, more importantly, what will be the scope of whatever interesting laws do involve those categories. It is the denial of this latter point that, I believe, is required to provide any motivation for an essentialist position and that, I have argued, is very difficult to reconcile with the range of phenomena I have been discussing.

Let me conclude with a word about natural kinds. There is certainly no harm in calling a set of objects that are found to have a substantial number of shared properties a natural kind. I want to insist that the discovery of such a kind provides no basis for the supposition that some particular property or properties can non-arbitrarily be singled out as essential. But, as I remarked earlier, there is no reason why the term 'natural kind' should be wedded to essentialism—or, anyway, no more reason than an accident of linguistic history that can readily be rectified. With this proviso, I am quite happy to refer to species as natural kinds. This case is unusual, in that we do have reason to expect that members of species will share a large number of properties, this reason being that we suppose the members of the species to have come about through an extremely homogeneous historical process. However, this in no way contradicts my insistence that the extent of homogeneity within a kind should be treated wholly empirically. Members of a species, as I have remarked, also vary greatly. And we cannot know a priori how variable any particular feature will turn out to be.

The only thing that could provide grounds for dispensing with this empirical stance would be if we were somehow to know that the members of certain kinds were completely homogeneous in all respects. Many people seem to believe that this is true of the kinds distinguished by physics and chemistry, though I find this doubtful.[17] If it is, physics and chemistry are in a very important respect different from biology. But even if this is the case, it is surely an empirical fact, not anything that could be known a priori. It is surely possible to conceive of a world composed of indivisible atoms, each as different from one another as one organism is from the next. Such would not appear to be the case; but *how* homogeneous physical or chemical particles may be remains an empirical matter. If this is correct, even microphysics cannot provide a hiding place from the categorial empiricism that I am advocating.

[17] I have briefly defended this claim elsewhere (Dupré 1983: 326–7).

9

What the Theory of Evolution Can't Tell Us

Let me begin by saying that I am a great enthusiast for evolutionary theory. Darwin's ideas and their subsequent development over the last 140 years are surely among the greatest insights into nature that humans have so far achieved. I make this perhaps perfunctory genuflection because scepticism about some of the more extreme claims made for the insights to be derived from evolutionary theory can incite accusations of insufficient respect for Darwin's achievement. In further exculpation, let me note that it seems to me a common sequel to great scientific insight that enthusiasts will exaggerate the range of its implications. No disrespect is implied to Newton's achievement by the observation that Laplacean visions of an entire universe to be understood in terms of the collisions of tiny billiard balls have proved to be an overstatement of his theory's capabilities. I believe that we are now seeing overzealous extensions of Darwin quite as egregious as Laplace's extrapolations of Newtonian theory.

Unsurprisingly, the area in which, I shall argue, the fruits of evolutionary theory are currently being most egregiously exaggerated is in the understanding of human behaviour. It will be best to approach the topic, however, by considering what the theory of evolution uncontroversially *can* tell us. Most obviously, it provides an account of how such complex and improbable-seeming entities as biological organisms could have come into being at all. Prior to the articulation of a plausible and broadly mechanistic answer to this question it was perhaps not unreasonable to suppose that the only possible solution to the problem was an appeal to an intelligent creator—though this solution is greatly vitiated by its obvious tendency to launch a vicious regress. The continued articulation of the Darwinian explanation —its extension to the earliest stages of chemical evolution and the detailed description of processes of evolutionary change, for example —remains a central and fascinating scientific project.

A more difficult issue is what evolutionary theory can tell us about particular kinds of organisms. Evolutionary theory is often thought to show us something about the functional analysis of organisms, the attribution of functions to their parts or behaviour. Indeed, philosophers have proposed redefining biological function in terms of the selective processes that are thought to have led to the development of a part or a behaviour, or, better, that have maintained its existence in recent evolutionary history. I have no objection to this idea, or this redefinition. An important proviso should be entered, however. Much functional knowledge pre-dated evolutionary theory and is in no way hostage to the claims of that theory. The function of an eye is to see; and it is a condition of adequacy on an evolutionary account of function that it recognize this fact. There are two ways in which evolution can modify or amplify such knowledge. First, we often discover that the reason an organ evolved may be quite different from the function it currently serves. Thus lungs, it is claimed, evolved as swim-bladders. But this does not begin to show that the function of lungs is to keep us afloat.[1] Similarly if eyes evolved, say, to mimic the markings on nasty-tasting little animals this does not begin to threaten the idea that the function of our eyes is to let us see.

Evolutionary theory may also help us to elucidate the function of organs or behaviours whose function is obscure, or explain the presence of organs or behaviours which appear to lack functions (they may have had some function in the past). Consider the classic example of the peacock's tail. It is initially puzzling why an organism would be encumbered with such an unwieldy piece of equipment, and one might even suppose that its only possible function was to provide ornamentation for Victorian ladies' millinery. Evolutionary theory has suggested an account that is now very widely accepted: the real function of the peacock's tail is to attract peahens; and to do this it must compete with the splendid appurtenances of other peacocks. Here also we see the connection of evolutionarily grounded accounts of function with more intuitive accounts. An organ has a function for an organism if it does something useful for it; attracting peahens is something to which peacocks generally attach a good deal of importance; and flashy tails are a sine qua non for achieving this goal. And of course a trait or a behaviour is useful for the organism, in the relevant sense, if it helps the organism to be treated favourably by natural selection. As a final comment on this possibility, to confirm that the function

---

[1] For a philosophical theory of biological functions defined explicitly in terms of recent evolutionary history rather than evolutionary origin, see Godfrey-Smith (1994).

of the peacock's tail is as suggested, we must confirm that peahens are, indeed, more willing to mate with peacocks with gaudy tails. If they are not, attracting peahens cannot be their function. Perhaps they once functioned to attract peahens, and thus evolved, but now have no function at all. A general theme of this essay is that the theoretical in biology must always be subordinate to the empirical.

As one more preliminary before approaching the topic of human behaviour, let me say something on the topic of optimality. Perhaps when it was believed that organisms were created from whole cloth (or whole protoplasm) by an omniscient and omnipotent designer it would only have been pious to assume that the design of organisms was optimal. Without qualification, such an assumption has no place in contemporary evolutionary thought. The point is nicely encapsulated in a little anecdote. Two anglers, so the story goes, are fishing for salmon in Alaska. Suddenly one of them sees a polar bear rapidly approaching. One of them hurriedly removes his waders and begins putting on his sneakers. 'You fool,' says the other, 'you can't outrun a bear whatever you're wearing.' '*You* fool,' says his companion. 'I don't have to outrun the bear. I only have to outrun you.' The point of this little episode of natural selection barely needs elaboration. Evolutionary theory has no place for a concept of absolute optimality. If optimality analysis has any use, it is only to distinguish among the range of features or behaviours actually available to a population of organisms. One does not need a jet engine to escape from predators; one just needs to be faster than one's fellows. This simple point should remind us, again, to be highly suspicious of a priori theorizing about the course of evolution. Design analysis of supposed evolutionary problems for hypothetical ancestral organisms has little significance in the absence of detailed knowledge of what design features were actually available to actual ancestors.

With these points in mind, let me proceed to human behaviour. My topic is the evolution of the human mind, but it will be better to begin with quite a different though no doubt a related thing, the human brain. We may say with some confidence, I think, that one of the functions of the human brain is to play a central role in the production of behaviour. I say one of the functions, as I shall not be concerned with vital but presently irrelevant functions such as the regulation of body temperature, heart rate, and so on. Experience of damaged brains leaves no doubt that the brain is causally necessary to the production of sometimes appropriate behaviour, and there can surely be no doubt that the tendency to produce some kind of evolutionarily successful behaviour must have been a necessary condition for the evolution of the remarkable cerebral hypertrophy of *Homo sapiens*. Large brains, it is sometimes suggested, may have been selected historically by runaway sexual selection rather than because they offered any great benefits for everyday survival. Whether or not this is true, large brains have enabled us to create societies the complexity of which calls for a fairly large brain for successful functioning. So whatever the evolutionary history of our brains, their ability to facilitate the complex behaviour required of modern humans is surely now part of their function.

While we now know a good deal about brains and a good deal about minds, we know almost nothing about how the properties of the former contribute to the functioning of the latter. For most of this paper I shall be concerned about minds. This is not the place to address the crude assimilation of mind and brain to which philosophers have sometimes been tempted. One point of difference should however be emphasized. Whereas there is no doubt that the brain, *qua* physical structure, is a product of biological evolution, it is much less clear to what extent this is true of the mind. The gross morphology of the human brain presumably depends for its ontogeny on complex features of the human genome, and these features presumably evolved over millions of years. The human mind, on the other hand, is arguably subject to much more rapid evolution and much greater developmental plasticity. The mind of a contemporary overeducated American or European academic appears to have developed along very different lines from that of a contemporary hunter-gatherer in New Guinea and probably even a contemporary rural American farmworker; all of these are surely a very different thing from that of a peasant in ancient Assyria. The mind develops through experience and education in interaction with a highly complex cultural context. And this cultural context is subject to enormously more rapid evolutionary change than are complex physiological structures.

It is a familiar and trivial point that whereas our brains surely evolved, much of the behaviour they now allow us to engage in is very different from any that could have served to further their evolution. While our brains were evolving, nobody was engaged in the study of quantum mechanics or the writing of symphonies or even, more closely connected to the mechanics of biological survival, shopping at the local supermarket. Clearly our brains do not limit us to those kinds of behaviour that may have served our interests in the Pleistocene, that period of human evolution so beloved of sociobiologists or, as they now prefer to be known, evolutionary psychologists. Nevertheless, these theorists insist that our brains did evolve in ways that severely constrain our present behaviour, ways that can be greatly illuminated by consideration of the exigencies of life in the Pleistocene. Central to this programme is the idea that the brain has a complex evolved structure that underlies behaviour equally today and for our ancestors dead these many millions of years. More specifically, it is argued that the brain is divided into many distinct components, or modules, each designed by evolution to perform optimally on a quite specific task. Since, as I have mentioned, the vast majority of human evolution is supposed to have occurred in the conditions of the Pleistocene, the canonical method for identifying these modules is to speculate on the demands that Pleistocene life would have imposed on early humans, and consider what bits of information-processing machinery in their heads would have best equipped them to respond to these demands. It is to such views that I now turn. My main focus will be evolutionary theories concerning human sexual relations, though I shall conclude with some brief comments on the problem of altruism. As we all did long ago, I shall begin with sex.

It is hardly surprising that sexual behaviour should be a matter of interest to evolutionary psychologists. From an evolutionary perspective, reproduction is the most significant thing that organisms do, and until quite recently sex was a necessary condition for human reproduction. Evolutionary psychologists do not disappoint: there is indeed a substantial industry of speculation about the evolutionary origins of human sexual behaviour. Before considering this in any detail, however, it is worth asking what kinds of things we might hope to learn from such an enterprise. To begin with a negative point, what we should not expect is to learn much about what kinds of sexual behaviour humans do in fact engage in. This should, I hope, be clear from my brief preliminary remarks: evolution does not predict the specific characteristics of organisms. Sexual behaviour, in particular, is enormously diverse across species, and is enormously diverse within our own phylogenetic lineage, the primates. It would seem, therefore, that there can be no hope of telling a priori what kinds of behaviour from among this range of possibilities humans will have come to exhibit, though many contemporary evolutionary psychologists do appear to entertain just such a hope.

A more plausible goal is explanation. Given the identification of characteristic human behaviour, we might hope to identify the selective advantage it confers on us, or conferred on our ancestors. This suggests a somewhat more ambitious idea. Prima facie, human sexual behaviour is extremely diverse. A number of philosophers have identified the fundamental feature of explanation as unification. Thus we might hope that underlying the apparent diversity or polymorphous perversity of human sexual behaviour we might find basic themes over which this apparent diversity provided only a superficial gloss. Finally, and much more controversially, one might expect evolutionary explanation of sexual behaviour to provide some kind of justification. Although evolutionary theorists almost universally deny any such goal, it is hard to see how they can avoid offering this possible use of their work. If evolution has in fact shaped our behaviour, it can only have done so by selecting physical structures, presumably in the brain, that cause the production of such behaviour. To say that a certain behaviour, which some find morally objectionable, is caused by a physical structure in my brain, is in effect to remove at least part of my responsibility for it. If my arm jerks convulsively and knocks your Ming vase to the floor, you will feel rather differently about my behaviour if you learn that I am subject to unpredictable brain spasms that cause my limbs to jerk wildly, than you would if you believed this to have been a freely chosen act of destruction. Similarly, if you are inclined to feel some moral disapproval at hearing that I have abandoned my wife and children and run off with a woman twenty-five years younger, your censure will surely be mitigated if you are convinced that there is a part of my brain the express function of which is to cause me to do just this. (It is also possible to make the opposite argument, condemning behaviour that has not evolved properly. While writing this paper, I happened to hear Rush Limbaugh informing the world that the reason homosexual marriage should be legally prohibited was that hundreds of millions of years of evolution had produced contemporary American matrimony, the best possible vehicle for human civilization. I'm relieved to discover, at least, that Mr Limbaugh is among the minority of Americans who believe that humans did, indeed, evolve.)

Human sexual behaviour is, as I have already remarked, prima facie diverse. People differ in the objects of their sexual desire (heterosexuality, homosexuality, and bisexuality, to mention only the most common), the frequency of their sexual behaviour, the kinds of sex acts they engage in, the number of partners they engage in sexual behaviour with, and no doubt much else besides. An obvious difficulty for the claim that sexual behaviour is much illuminated by evolutionary theory will be to account for the persistence of such diversity. Why, that is, has evolution failed to select more reliably the biologically fittest from among this range of options? The most obvious problem is presented by homosexuality. Although discussions of this topic seldom bother to offer empirical data as to whether homosexuals really do have lower fertility than heterosexuals, it is fairly uncontroversial that sexual preference for members of the same sex is an improbable strategy for increasing reproductive success. Even leaving aside this problem, variations among such things as numbers of sexual partners and age preferences for mates suggest that sexual behaviour is not greatly constrained by evolved biological forces.

The standard response, once again, is retreat to the Pleistocene. Evolutionary psychologists are accustomed to note that the conditions of modern society have existed for no more than the blink of a geological eye. This eyeblink is generally supposed to have been preceded by a million years of relatively constant conditions for early humans as hunter-gatherers on a proto-Savannah. Perhaps, then, in this evolutionary past, sexual behaviour was strongly selected and consequently much more homogeneous. The dispositions evolved at that time, it may be said, still exist, and exist in purpose-built mental modules; but in modern societies they come into conflict with various and diverse cultural forces. The interactions between these and the evolved biological dispositions might be diverse and even partially indeterminate, but it might still be possible to extract the biological commonalities by careful analysis, ideally cross-cultural, of the existing range of sexual behaviour. Some picture like this seems to be assumed by much current work in the evolutionary psychology of sex. Two obvious problems suggest themselves. First, how could we come to detect these alleged biological dispositions through the noise of cultural contamination; and, second, what would follow for our understanding of actual sexual behaviour if, somehow, we did?

A topic with which evolutionary psychologists have been much exercised recently is that of sexual attractiveness. Obviously enough, it is of great evolutionary importance to select suitable mating partners. In arguing for the necessity of domain-specific psychological mechanisms, Donald Symons, a leading luminary in this area, notes that a koala cannot use the same criteria for selecting food and selecting mates: attempting to mate with eucalyptus leaves is a quick evolutionary dead end. The necessary discriminations are more subtle even than this. Another expert on the topic, Bruce Ellis, explains:

Consider . . . a woman who can choose between two husbands, A and B. Husband A is young healthy, strong, successful, well-liked by his peers, and willing and able to protect and provide for her children; Husband B is old, weak, diseased, subordinate to other men, and unwilling and unable to protect and provide for her and her children. If she can raise more viable children with husband A than husband B, then his mate value can be said to be higher. (Ellis 1992: 267)

One might suppose that the evolutionarily correct solution to this problem would be simple enough. But as the editors of the anthology in which this article appears remind us, with specific reference to such seemingly 'simple' activities as finding someone beautiful or falling in love, 'our phenomenal experience of an activity as "easy" or "natural" can lead us to grossly underestimate the complexity of the processes that make it possible' (Barkow *et al.* 1992: 245).

Let us consider, then, the problem of mate choice facing the human male. A human male intent on reproductive success must not only distinguish a female of the same species, but it is better if she be post-pubescent, pre-menopausal, and non-pregnant. At this point evolutionary psychologists appeal to a standard quasi-economic model. Males, they argue, are evolutionarily designed to select females with the maximum fertility. Females, requiring little actual biological contribution, seek males with the ability to invest most heavily in the raising of offspring (and, if possible, with some indications that they will be inclined to do so). Focusing for now on the problem for males, various criteria have been suggested for identifying suitable mates, ranging from the fairly obvious, such as the absence of visible lesions or open sores indicative of inferior health, to the highly esoteric, such as an ideal measure of waist-to-hip circumference (around 0.7), a suggestion that has been the subject of considerable research (see Singh 1993). General agreement centres on one point, however: males will be most attracted to females at the beginning of the fertile stage of the life-cycle, and who have not yet borne children. For modern humans this suggests a peak attractiveness in the range of 15 to 18 years of age. Donald Symons predicts that in every society women are considered more attractive at 18 than 38 (Barkow *et al.* 1992: 143). Let us consider this last proposition in a bit more detail.

It is worth noting briefly the obvious complication that a 38-year-old Michelle Pfeiffer, say, might generally be judged more attractive than even an averagely attractive 18-year-old. Presumably the most that could be claimed was that 18-year-olds were more attractive than 38-year-olds other things being equal, whatever exactly that means in this case, though if my intuitions are correct here common standards of attractiveness are by no means strongly correlated with judgements of potential fertility. Moreover, the theory says nothing about why, even among 18-year-olds or among 38-year-olds, small differences in the arrangement of facial features are commonly taken to ground substantial differences in physical attractiveness. It doesn't seem very plausible that somewhat closely spaced eyes, say, or a rather large nose, are features much correlated with potential fertility. It would also be interesting to know, in passing, whether Calvin Klein is succeeding in selling clothing on the evident, but sociobiologically questionable, assumption that seriously undernourished 13-year-olds realize an ideal of female beauty. Letting all this pass, let us assume for the sake of argument that there is indeed some male psychological mechanism for assessing female attractiveness, and that this is maximally aroused by 18-year-old females. Let us even assume that among our Pleistocene ancestors—stereotypic cavemen—attractiveness in this sense was the sole grounds of mate-selection. Two qualifications are surely indisputable. First, mate-selection among modern humans is enormously more complex than it was among these hypothetical cavemen. I suppose that so-called trophy wives would not be accounted trophies if there were not some recognized virtue to mere youthful good looks; but a trophy wife seriously deficient in intelligence, charm, good manners, and so on would, I assume, be more often an embarrassment than a prize. Prudent mate-selection, that is to say, involves a wide range of factors, many of which have nothing whatever to do with purely physical attractiveness. Though it is certainly a familiar phenomenon that people, especially men, will sometimes select a mate solely on the basis of physical attractiveness, this is almost universally regarded as highly imprudent.

The second, and perhaps even more important, qualification is that mate-selection, in the sense of selection of a long-term partner for the bearing and rearing of children, is hardly the sole context in which modern humans make judgements about the attractiveness of other people. Whether or not this was true of our less sophisticated ancestors, contemporary humans are interested at different times in a variety of different kinds of relationships with members of the opposite sex (or, if they are homosexual, the same sex; though how this relates to the present issue is entirely unclear). They may seek friendship, casual sex, a brief romance, lifelong companionship, a co-parent for their children (existing or yet to be born), a status symbol, a domestic drudge, and so on. Presumably the relevance of prehistoric whisperings concerning reproductive potential will vary considerably from one to another of these cases. This, in turn, suggests a very general problem for the allegedly modular evolved mind. However modular the mind may be, the output of such modules must somehow be integrated into some broader process in which whole human beings come to make decisions, and must be capable of weighing modular outputs differently according to different ends to which the decision-making process may at any time be directed. There must be some part of the mind in which it is possible to decide whether to pursue a potential mate or stop for dinner. Suppose, for the sake of argument, that there is indeed a mechanism in the human brain that disposes men to select very young women or girls as ideal mates. Given that this atavistic mechanism provides only one of a range of inputs into actual processes of mate-selection, and given that mate-selection, in the sense assumed by evolutionists, is only one of a range of kinds of behaviour in which this hypothetical machinery might figure, it is not at all clear that identifying such machinery will tell us anything much about the behaviour or even behavioural dispositions of modern humans.

It seems to me, then, quite unclear what would follow for our understanding of human behaviour if we did successfully identify these atavistic mental modules. Perhaps the most plausible answer is that we would learn something about psychopathology: the maladapted mind, the mind unable to function in the conditions in which it finds itself, is perhaps a mind constantly and uncontrollably driven by uninterpreted urges from its evolutionary past. At any rate, whether or not much would follow from the identification of these modules, I want now to look at the processes by which they are, allegedly, disclosed. (And clearly, if nothing much follows from their existence, they will be that much harder to detect.)

Whereas in the early days of sociobiology a priori argument and stereotyped clichés about sexual difference were generally considered sufficient evidence for the evolutionary illumination of human sexuality, recently these have been augmented by a great deal of empirical study. Whether much of this research provides compelling evidence for the hypothesized psychological mechanisms is another matter. In some cases the research involved would most naturally be taken as self-parodic farce, if it were not in fact taken as serious science, published in leading scientific journals. I think, for instance, of the investigation of the hypothesis that rape is an evolved male reproductive strategy by the method of so-called objective phallometry, which involved the exposure of prison inmates to filmed depictions of rapes, with measuring devices attached to their penises (see Thornhill and Thornhill 1992). Even discounting the possibly atypical nature of this sample, the inference from sexual arousal by these movies to an innate disposition to rape has all the plausibility of the inference that overweight middle-aged men who show objective signs of excitement when watching football games on Sunday afternoons must have some disposition to play professional football.

Somewhat less absurdly, David Buss has investigated the criteria of mate-selection by sending questionnaires to substantial numbers of people in a variety of cultures asking them to rate the importance of various characteristics for potential mates (Buss 1994). He did indeed discover that in most cultures men claim to attach more significance than women to youth and beauty. On the other hand the significance of chastity, which evolutionary psychologists predict should be of great importance to men but much less to women, was found to vary greatly from culture to culture. In China, for example, it was found to be of great and equal importance to both sexes, in Germany or Sweden it was of very little importance to either. Thus Buss concludes that 'some preference mechanisms are highly sensitive to cultural, ecological, or mating conditions, while others transcend these differences in context' (Buss 1994: 254). It is, of course, equally possible that the cultural circumstances that encourage some of these preferences are currently less variable than those that support others. Buss also found, as will surprise no one but an evolutionary psychologist, that two characteristics consistently valued by both sexes were kindness and understanding (a portmanteau property apparently), and intelligence. Wonderfully displaying the accommodating nature of the programme, Buss proposes that 'kindness ... may provide a powerful cue to the willingness of a man to devote resources to a woman and of a female to devote her reproductive resources to a particular man' (Buss 1994: 255). Possibly so, though it might also lead the man to bestow his resources on others, or contribute to charity, or, to invoke a counter-stereotype, the whore with a heart of gold, it might even induce the woman to bestow her reproductive resources more liberally. The rather banal explanation that it might be pleasanter to spend one's life in the company of a kind and understanding person also has something to commend it. Similarly, without dismissing out of hand the possibility that intelligence might be a cue for superior hunting ability, ability to evade large predators or distinguish edible roots, and so on, the plausibility of the consideration that life with someone dull and witless might be less rewarding renders these more exotic speculations somewhat redundant.

The ease with which evolutionary psychologists can accommodate data is even more striking in a paper by Bruce Ellis commenting on the fact that in questionnaires women claimed to attach little importance to either dominance or social status. Ellis offers four possible explanations: they may mistakenly have supposed that the men were disposed to dominate them rather than other men; they may be reluctant to admit that they prefer such men; they may prefer such men but be unconscious of the preference; or their assumed reference class may only include high-status men, among whom details of status will not be important (Barkow *et al.* 1992: 282). Perhaps so. But methodologists will surely see here the stigmata of a research programme with a whiff of degeneration—if, indeed, that does not imply more antecedent progressiveness than is evident.

A different experimental methodology involves showing subjects pictures of members of the opposite sex and asking them how attractive they found the persons depicted, or whether they would be willing to engage in various kinds of relationship with them. Such procedures were used, for example, in the investigation of the hypothesis that men were attracted to women with a 0.7 waist-to-hip ratio. One should of course note at the outset that such experiments presuppose an extremely narrow and simplistic conception of attractiveness. It is difficult to make reliable judgements from a still photo as to kindness and understanding, or intelligence, for example. Nevertheless, supposedly supporting the hypothesis that women seek higher status in men, it was duly discovered that female undergraduates and law students were more willing to engage in various relationships with a 'homely' man in a designer blazer and Rolex watch described as a physician than with a 'handsome' man dressed as a Burger King employee. In opposition to some obvious rival explanations of these phenomena, researchers infer that even very high-status women seek male partners of even higher status (Buss 1994: 101).[2] Men, by contrast, were said always to prefer the handsome to the homely women. All I shall say here is that the pressure to assume a purpose-built psychological mechanism to explain such results is, to say the least, resistible. One might only note that Americans, we are often told, spend an average of six or seven hours a day watching television. A large part of their experience is thus of people leading lives far more packed than their own with excitement, adventure, sex, and romance. The men in this vicariously experienced world are intelligent, powerful, well-dressed, and of high social status; the women are young and beautiful. The men may be old: no female romantic heroines are of the age of Sean Connery or Clint Eastwood. The men are also handsome, which might explain the somewhat embarrassing fact for evolutionary psychology that women are found to prefer handsome to homely men even if this is not so decisive a factor as it appears to be for men. I make no claim as to how far this does explain the phenomena in question, and the direction of causation could be debated endlessly. I do claim, however, that such experimental results do nothing whatever to sort out the alleged psychological mechanisms from obvious candidate cultural influences.

This empirical research is, then, disappointing. Various hypotheses are suggested for the criteria of human mate-choice or judgement of attractiveness. Many of these are familiar and banal. Sometimes, especially in the banal cases, the hypotheses are to some degree confirmed. This shows that the psychological mechanism is at work. In other cases they are not. This with the help of a range of ready-to-hand auxiliary hypotheses shows that the psychological mechanisms interact with cultural forces. It would be not only grossly premature, but potentially dangerous, to suppose that much had been learned about the innate bases for human mate-choice.

Let me now insist, in spite of the preceding scepticism, that I do not deny that behaviour must be seen as conditioned by both cultural and other environmental forces and by human psychology. And culture itself must in significant ways be shaped by human psychology. Indeed I do not even want to deny that at some level there

---

[2] My own informal research on this question has found high-status women claiming a strong preference for the hunk from Burger King. I must confess, however, that my sample size is on the small side.

is such a thing as human nature. Experience and intuition confirm, for instance, that humans do not thrive under conditions of slavery, a fact that would be difficult to understand if humans were mere blank slates on which any set of behavioural dispositions could freely be written. But the way to deepen our understanding of human nature is not by a priori evolutionary theorizing, any more than this is the way to deepen our understanding of physiology. I want now to suggest how my criticisms of neo-sociobiology might connect with some more theoretical issues about the nature of evolutionary processes.

I do not propose to deny that the mind may contain task-specific mental modules, nor to affirm it. Although evolutionary psychologists and others offer broad a priori arguments in favour of their modules I do not find these especially convincing. As has been a theme of this paper, I think we should avoid a priori arguments throughout biology, and I doubt whether there is any compelling evidence currently to be had on this topic. But while I remain agnostic about these hypothetical modules, I am more strongly sceptical of the atavistic character attributed by evolutionary psychologists to them—the assumption, that is, that we are behaviourally programmed for the life of hunter-gatherers on the Pleistocene savannah. (Newt Gingrich was clearly in accord with the latest scientific thinking when he spoke of the universal need of men to go out and kill giraffes.)

It is not difficult to understand the intellectual origins of this atavism. Mental modules are taken to be, or at least to be grounded in, specialized structures in the brain. Specialized physiological structures, it is generally supposed, can only be built by complex genetic programmes. And complex genetic programmes take hundreds of thousands or millions of years to evolve. Certainly they cannot have undergone significant evolutionary change in the few millennia of modern human existence, and thus atavism is an inevitable concomitant of genetically evolved behavioural dispositions. So, although evolutionary psychologists do not deny the obvious truth that behaviour depends on both evolved neural structures and environmental stimuli, they insist on major asymmetries between these. Clearly evolutionary psychologists must concede that an empirical question remains as to how flexible these structures will be in response to environmental circumstances, and an empirical question with widely differing answers: men can learn to tolerate the promiscuity of their mates, for example; but nothing, we are told, appears to shake their desire for the young and the beautiful. Although there is certainly nothing surprising in the idea that a mental system should respond differentially to variations in its environment—presumably that is an essential aspect of being more or less intelligent—from the perspective of evolutionary psychology it is surprising good fortune if the range of those responses extends to behaviour appropriate to modern circumstances. It is, in fact, at the core of the whole project of explaining behaviour in terms of evolved mental modules that behaviour should be seen as always hovering on the brink of contemporarily irrelevant atavistic tropisms.

Let me introduce some fundamental worries about this project with a quote from the front page of a recent campus newspaper.[3] Describing the use of computers by scientists at the Stanford Medical Center to assemble maps of the approximately 100,000 genes on the human chromosomes, the report notes that 'This collection of genes contains the complete instructions about how to make a human body.' Something like this is, I suppose, widely held to be a contemporary truism, and a truism that underlies the assumption that the structure of our brains is more or less determined by features of our chromosomes. Yet far from being a truism, this statement is completely and hopelessly false. Leaving aside the question whether the notion of 'instructions' is even intelligible outside the context of a symbolic system, we should note that a full set of human chromosomes has no chance of producing a human body unless it is situated in a human cell replete with all the proper transcription, energy-processing, etc. cellular machinery; that the cell must be properly situated in the right part of a human female body; and that the human female must be situated in an adequately sustaining social and ecological environment, and so on. It is often supposed—and here Richard Dawkins has had a pernicious influence—that there is a fundamental asymmetry between the genes on the one hand and all these contextual prerequisites on the other. Probably this is most commonly expressed in the idea that the genes carry information for which these various levels of environmental context provide merely a channel. But this is an untenable position, a realization that has been made clear by the work of developmental systems theorists (see Oyama 1985). Simply put, there is no sense of information in which the genes carry information that can be read off by the environment in which it cannot equally be said that the environment carries information which is read

[3] *Stanford Report*, 22 May 1996.

off by the genes. More simply still, there are many sources of information that are required in building a human body—genetic, cellular, physiological, and, especially at later stages, cultural. All are necessary, none is sufficient.

Returning to the main topic, then, let us assume that among the characteristics of humans are mate-selection mechanisms. Leaving aside the worry mentioned earlier about the variety of activities that seem to be grouped together under this rubric, we might at least agree that some or all of these activities are of great importance to most humans. How are these mechanisms transmitted, to whatever extent they are, across generations? We have not the slightest idea. There may well be a major genetic component in such transmission, though the observed intercultural variations in, and intracultural rates of change in, the relevant behaviour strongly suggest that more rapidly variable elements, presumably cultural, are also centrally involved. In the absence of the genocentric picture that I have been criticizing, there is, at any rate, no reason to assume that the mode of transmission is genetic and, therefore, no reason to assume that our behaviour is adapted to the conditions of our distant past.

I have so far taken it for granted that it is clear what is at stake in claims about the modularity of mind. But in reality this is anything but clear. No one, I assume, thinks the brain is a wholly amorphous structureless mass, and no one supposes that it is exhaustively divided into wholly independent systems entirely dedicated to particular areas of behaviour. The reality, I think, is that claims about mental modules are claims that fairly specific aspects of behaviour are generated by genetically coded structures in the brain. Although there are perhaps no strict genetic determinists—the causal chain from gene to behaviour is too long and attenuated for such a view to be remotely plausible— specialized evolved mental modules are nevertheless the contemporary vehicle through which somewhat qualified versions of genetic determinism are currently expressed. With arguments for the existence of such mental modules almost exclusively aprioristic, and evidence for the behaviour they are supposed to generate equivocal at the very best, we should treat this school of genetic determinism with all the respect that thinking people have come to accord to its predecessors.

Let me conclude with a few remarks about an equally interesting if less sexy topic, altruism. Altruism has always been perceived as something of a problem by evolutionists (often referred to, indeed, as the problem of altruism). This is the problem of organisms helping other organisms at some net cost to their biological fitness, something apparently precluded by evolutionary theory, but nevertheless occasionally, at least, observed.

One reason that the problem is so intriguing is that there is something perverse, almost, in the very idea of a 'problem of altruism'. One might be forgiven for supposing that there was at least a more pressing 'problem of individualism'. Individualism is perhaps the dominant intellectual idea of the twentieth century. Political individualism, we are often told, has just emerged triumphantly from an epic battle with its nemesis, socialism. Methodological individualism, as realized in neoclassical economics, game theory, and the various manifestations of rational choice theory, increasingly dominates the social sciences. Despite flirtations with behaviourism and instrumentalism, this methodological individualism is generally grounded in ontological assumptions about human nature. As David Gauthier has expressed it, 'beyond the ties of blood and friendship . . . human beings exhibit little positive fellow feeling' (Gauthier 1986: 101). The common grounding, finally, for this ontological belief is in evolutionary biology. Indeed, Gauthier's references to 'blood and friendship' may naturally be taken to refer to kin altruism and reciprocal altruism, the two circumstances generally allowed by evolutionists to provide exceptions to the rule that evolution promotes an exclusive attention to self-interest.

The problem of altruism, then, is the problem of how, given this pyramid of individualist ideology, altruism is possible at all. At the level of economics altruism appears simply as a somewhat eccentric taste that some people exhibit. And *De gustibus non disputandum est.* But at the more fundamental level of biology it is a problem that continues to be grappled with. The very connection with this problematic ideological pyramid should encourage us to look very closely at the assumptions about evolution that present altruism, that is, the failure of individualism, as such a problem.

The basic picture at issue, of course, is of evolution as a deathly struggle for survival between individuals. Supposedly this 'tough-minded' picture was carried one step further by Dawkins's retreat to the allegedly more fundamental selfish gene. Ironically, this move suggests from another point of view a hint of desperation. For, of course, genes are about as massively cooperative a set of entities as one could hope to find; one doesn't provide 'instructions for making a human body' or even make a major contribution to such without working together a bit. What this illustrates is that competition, or any internally integrated behaviour at one level, requires massive cooperation at others. Thus, finally, the alleged omnipresence of individualism makes sense only if the individual is the only significant level at which selection has shaped the human.

Recent thinking about evolution makes this highly implausible. First, it is becoming increasingly orthodox to recognize that selection may occur simultaneously at many levels both above and below the individual. Second, the very notion of a biological individual as an intuitively obvious entity has come into question. In part, this has come about through consideration of clones, social insects as super-organisms, obligate symbiosis, and so on. Developmental systems theory, the perspective to which I alluded briefly in my brief remarks about genocentrism, suggests that one should see in evolution a number of overlapping and interacting patterns of self-replication that may coincide neither in time nor level of organization. If this is right, we should expect the prevalence of individualistic and altruistic tendencies at any particular level to be a strictly empirical question. All this fuzziness will be anathema to those who like their science clean and crystalline. However, returning to the problem of altruism, one thing in favour of the fuzzy is its traditional association with the warm.

VI

Differences between Humans
and Other Animals

10

The Mental Lives of Non-Human Animals

Introduction

It is commonly supposed that the question whether animals other than ourselves have minds is perfectly simple to understand, but very difficult to answer. We suppose that we know exactly what is at issue, since we know from our own experience what it is to think, or more generally have mental lives; but we are very uncertain how we might ever discover whether animals do the same. In large part, we may also suppose, this is because, being dumb, they are unable to tell us. I want to argue in this paper that almost everything about this set of views is wrong. Our difficulty with this question is hardly at all to do with lack of evidence, but has everything to do with a lack of clarity about what is really involved in the attribution of mental states. I do not, of course, mean that the question about animal minds could be settled independently of any evidence. But I do want to suggest that the empirical facts in question may, in many cases, be quite banal. The trick is to decide what the relevant facts are. To put this claim in an imperialistic mode, I am suggesting that the problem is paradigmatically philosophical.

In the first part of this paper I shall expand on the preceding claims and explain in a general way the kinds of questions I do take to be involved in deciding whether an entity is an appropriate subject for the attribution of mental properties. In the second section I shall make some rough and tentative suggestions about the appropriateness of attributions of more specific kinds of mental phenomena. (One important moral of the opening section is that we should avoid assuming that there is some unitary answer to the question about

I have been helped by comments on an ancestral version of this paper by audiences at the University of Colorado at Boulder and the University of California at San Diego. I have also benefited greatly from the responses of various people to a previous draft. I am particularly grateful to Marc Bekoff, Dale Jamieson, Peter Hacker, Thomas Kuhn, and Harriet Ritvo.

animal minds.) In the final section I shall try to identify some questions that do remain mysterious, and offer some ideas about how light might be thrown upon them.

I would like to develop two related themes about the mental states of animals. First, I want to point out the powerful and pernicious influence that Cartesian assumptions, generally—if perhaps ironically —unconscious, continue to exert on much of our thinking about this topic, and to say something of what are the consequences of rejecting these assumptions. Cartesian perspectives are omnipresent in recent discussions of animal minds, not least among those most vocal in support of the view that non-humans have a wide range of mental capacities. One conspicuous example is the work of Donald Griffin (1984). It is even commonplace among those active in defending the rights of animals to ethical treatment. Marian Stamp Dawkins (1990), for example, rests the case that animals can suffer explicitly on the argument from analogy (for discussion of this argument, see below; see also Crisp 1990). My arguments against these Cartesian assumptions owe a great deal to my understanding of the later Wittgenstein (1953); a specific question I would like to explore is to what extent Wittgenstein's insights into the conceptual status of the mental justify us in espousing a kind of behaviourism.

Second, I want to argue against an idea, again more often implicit than explicit, that there is just one fundamental question as to whether animals *really* think. A close parallel here can be drawn with the question whether there are really animal languages, or whether animals can be taught *real* languages. In either case the question can be seen to presuppose a kind of essentialism, the view, that is, that there is some one crucial feature, an *essence*, that is necessary and sufficient to make a thing or phenomenon what it is. The questions then can be raised, What is the *essence* of thought (or language), and do animals have it? The idea that there is an essence of thinking is of course famously connected with the name of Descartes; its denial—and again the same denial for the case of language—is central to the work of the later Wittgenstein. So it should be clear that the issues I have distinguished have much to do with one another.

Descartes thought that the property that distinguished any genuinely mental phenomenon from anything else was its transparency, or indubitability, to the agent experiencing it. This is also the basis for a classical conception of consciousness. Objects of consciousness, on this view, are not only immediately apparent to their subjects, but their nature is unmistakable. Unlike Descartes, more recent thinkers do not necessarily identify mental phenomena with objects of consciousness. I say 'mental phenomena', though Descartes would say 'thought', because as suggested above one problem Descartes has bequeathed us is much too homogeneous a view of the mental. (In the *Meditations*, Descartes gives a typical list of kinds of thinking as doubting, understanding, affirming, denying, willing, refusing, imagining, and sensing; Descartes 1967: 153, Med. II.) Something is a pain, say, for Descartes, if and only if it is experienced as a pain by its subject; the nature of pain is unmistakably evident to the subject; and the meaning of the word 'pain' is exhausted by its function of referring to that experience. Although Descartes did not consider the minds of animals, presumably because he explicitly considered them to be nothing but machines, this conception of the mental raised a pressing problem about the existence of other human minds that has been prominent in the subsequent history of Western philosophy. It will be useful to approach the problem about other species by looking first at this problem about other members of our own species.

The relevant difficulty to which Descartes's conception gives rise is that the putative essential property of mental phenomena, transparency or consciousness, is accessible to us only in our own case. We cannot—and this is a logical rather than an empirical impossibility —ever have access to this property in the case of another mind. If, even in principle, we can never verify that the essential property of the mental is present in any case but our own, it is natural to ask whether there is any possible justification for believing in the existence of minds other than our own. The traditional answer to this question has been to appeal to what has become known as the argument from analogy. We observe in our own case, according to this argument, that certain of our mental states are correlated with characteristic modes of behaviour. We observe in other people these same patterns of behaviour, and infer inductively that they are accompanied by the same mental states (see also Crisp 1990).

It is worth remarking at this point that *if* this is the right way to think about other humans, there is little difficulty about other animals. The behaviour of my cat when it has a pain in its paw is very much like mine when I have a pain in my foot; on the other hand it never produces the behaviour which for me would be associated with extracting a square root. I would conclude, using the argument from analogy, that it had some but not all of the mental states I experience myself. The trouble is that if such an argument is needed, it is woefully inadequate. An inductive argument based on observation of one case to a generalization over a population of billions is hardly deserving of the title 'argument'. The reason that we do not accept inductive arguments based on a single instance is that we cannot, in general, have any reason to suppose that the observed case is typical. One would be in error, for example, in concluding on encountering a radio that all hard rectangular objects emitted complex and cacophonous sounds. In the present case, the very point at issue is whether consciousness, the property at issue in the argument from analogy, is a property peculiar to myself or more widely distributed. If the former, sceptical hypothesis is correct then the inductive argument—the argument from analogy—is worthless; to accept it is thus to beg the sceptical question entirely.

A radically different solution is offered by analytic behaviourism. According to this view, pain, for example, just *is* the characteristic set of its behavioural manifestations (see Skinner 1953; a rather more subtle position, though basically behaviourist, can be found in Ryle 1949). Thus to ask whether someone writhing on the ground with a knife in his leg is in pain is nonsense; nonsense of the same kind as the question, I know she has the same parents as I do, but is she really my sister? But attractive though this solution has sometimes appeared, it is obviously unacceptable in this simple form. It is possible that the person on the floor has his leg anaesthetized, and is pretending to be in pain. Given this possibility, it cannot be *nonsense* to ask whether the person really is in pain. It is worth noting again that *if* this were an adequate solution to the problem, it too would present no special difficulties of extension to other animals: if they produce the appropriate behaviour, they have the mental state in question, if not, not.

Since I cannot treat the ramifications of this problem in adequate depth in the context of this paper, I must now be somewhat dogmatic. My own view is that although, for the reason just stated, analytic behaviourism is untenable, there is a good deal right about behaviourism when it is separated from the claim that mental terms can be analytically reduced to sets of behaviours. What I take to be at the heart of Wittgenstein's attack on the Cartesian tradition is the demonstration that there are deep conceptual connections between mental states and the behaviour that constitutes their characteristic display. In differentiating this position from analytic behaviourism we must emphasize that a conceptual connection here must be distinguished from a strictly logical connection. The word 'pain' is not logically equivalent to some complex description of behaviour.

How, then, should we understand the meaning of a word such as 'pain'? Wittgenstein approaches meaning through a consideration of what it is to explain meaning. Meaning is what one grasps when one correctly understands an explanation of meaning. Meanings can be explained in many ways, one, but by no means the only, of which is the use of samples of the referent of a word (so-called ostensive definitions). So in the present context we need to think of what is involved in explaining the meanings of mental terms; and an essential part of the answer, surely, is behaviour expressive of the mental states in question. Moreover, as I shall elaborate in a moment, Wittgenstein shows that, contrary to the Cartesian picture, mental terms could not possibly be explained ostensively, i.e. by pointing to the connection between the word and the alleged mental referent.

The moral to be drawn here is not that mental states can be reduced to behaviour, but that, contrary to the Cartesian assumption, it must be possible to explain the meaning of mental terms through appeal to behaviour (or perhaps behaviour plus characteristic causal antecedents). Any explanation, for Wittgenstein, is fallible; it may be misunderstood. However, it must also be possible, if it is a satisfactory explanation, for it to be understood correctly. If it is, then the explainee has acquired criteria for the application of a term. Thus, for example, by observing myself and others in various painful positions, and producing various forms of behaviour expressive of pain, I can be taught the meaning of the word 'pain'. I can then apply it correctly to other cases. The distinction between Wittgenstein's position and that of analytic behaviourism, as well as further insight into Wittgenstein's positive view, can be found in his remark that the word 'pain' does not describe pain behaviour, but replaces it (1953: 89). Verbal expressions of pain thus become criteria of pain on a par with groaning; and like other criteria they can, on occasion, be disingenuous.

We can now consider the application of this picture to the worry about other minds, whether human or otherwise. I see a person with a nail stuck in his foot groaning and writhing on the ground. Since I am familiar with the criteria for pain, amply realized in the present case, this strikes me as a clear instance of pain. But there remains a philosophical inclination to ask: but is the person really in pain? Suitably expanded, this makes perfectly good sense. I might be asking, is this only a pretence of pain? Perhaps it is a rehearsal of part of a play, for instance. The legitimacy of such questions points exactly to the impossibility of providing a behaviourist reduction of pain. But of course the sceptic about other minds is not someone who wonders whether other people are only pretending to have the experiences they seem to be having. She is someone who wonders whether, in any case but her own, the situation observed really provides evidence of pain at all. But she is thus raising the question whether the criteria that she (thinks she) has learned for the term 'pain' really are such criteria. And this is equivalent to the question whether she really knows what it is that she wonders whether she is observing an instance of. The sceptical question thus appears to be self-defeating. And this, finally, suggests that the Cartesian perspective from which it derives must be confused.

This line of argument, though I believe it to be impeccable, seldom convinces. The reason is, I think, that most of us have retained extremely strong Cartesian intuitions. We think of our use of the term 'pain' as fundamentally a device for referring to something we are acquainted with in our own private experience, and thus as only contingently related to its typical causes and behavioural manifestations. To address this worry directly, we must turn to the most notorious and controversial part of Wittgenstein's overall strategy, the so-called private language argument. The Cartesian picture, as I have said, assumes that words for mental states function primarily to refer to private internal states; and if the function of a word is primarily referential, it should be possible to explain its use ostensively, by the use of samples of its referent. Thus, just as we explain colour words by using objects of the appropriate colour, we should explain sensation words by using samples of the appropriate mental quality. And to cut a long story short, this simple Cartesian picture of the meaning of a term such as 'pain' is incoherent. No one could explain the use of a term which functioned simply to refer to a private internal state; no criterion could be communicated for the application of the term, since only the person attempting to explain the term has any access to the private feature that is supposed to serve as a criterion; and hence no distinction could be imparted between correct and incorrect use of the term. This last point is especially crucial. Imparting a distinction between correct and incorrect application of a term is precisely what, as Wittgenstein emphasizes, an explanation of meaning is intended to achieve. So we have reached, in essence, the same point that was arrived at when the sceptical argument discussed above was argued to be self-refuting. The sceptical worry, which derives from this mistaken assumption about the meanings of mental terms, is raised in a way that undermines the meaningfulness of the very question that it is intended to raise.

This leads me, finally, back to the minds of animals. What I want to claim is that the starting point for an adequate approach to this problem is to reject the idea that there is some one 'deep' question involved. There is not, in Gilbert Ryle's (1949) memorable figure, some private internal stage across which the referents of mental terms act out their ghostly roles. Since we have no such stages ourselves, we need not enquire whether other creatures possess or lack them. We cannot even begin to consider a range of questions that continue to figure prominently in discussions of animal minds—Are animals really conscious? Are they self-conscious? Do they really know what they are doing? Do they have experiences? Or are they, on the contrary, merely machines?—unless we reject the Cartesian picture, and then ask what are really the criteria for the application of these various terms. When we do so it is, of course, likely that the questions will have rather various answers. In the next section I shall look at some questions about animal minds which, I think, can be answered fairly confidently.

Some Easy Questions about Animal Minds

In this section I shall touch on a number of different questions about the mental capacities of animals: Do they have experiences, beliefs, intelligence, or language? But I shall start with the most general such question: Are non-human animals ever in any kinds of mental states at all? As may well have been guessed from the preceding section, I take the answer to this question to be affirmative. However, I want to look briefly at some criteria that have occasionally been deployed to rule out any attribution of mentality to animals whatsoever.

Perhaps the commonest such criterion remains that of Descartes: Are animals conscious? This can be interpreted variously. Donald Griffin, in his book *Animal Thinking* (Griffin 1984: 9), suggests that the issue separating him from more behaviourally minded scientists is 'whether animals are mechanisms and nothing more . . . complex mechanisms to be sure, but unthinking robots nonetheless'. Elsewhere he suggests that the problem may be whether animals are aware of their own mental states (or: do they know what they are doing?), or again whether they are aware of objects not immediately present to their senses. (I shall return to this last idea.) The first question suggests exactly the aspect of Descartes's view that raises the insoluble problem of other minds: Whether an entity is conscious might be totally independent of the totality of its behaviour and behavioural dispositions. Although there are powerful intuitions in favour of such a possibility, I shall reject it for the reasons outlined in the preceding section: In the absence of behavioural criteria we cannot even attach a meaning to a mental term; so the attribution of mentality cannot be quite independent of behaviour.

Awareness of one's mental states, interpreted in one of Griffin's senses as 'knowing what one is doing', may seem closer to a genuine and empirical notion of consciousness. We sometimes distinguish, for example, conscious from unconscious mental processes precisely on the grounds of whether the subject is aware of what she is doing. A person acting on a post-hypnotic suggestion may offer an explanation of what she is doing in mentalistic or psychological terms, but the explanation may be quite mistaken. Unfortunately, it is pretty clear that this provides no adequate model for explicating the philosophical question about the mental states of animals. A person under post-hypnotic suggestion typically *is* aware of mental states; it is just that she is deluded about what mental states are in fact relevant to explaining her actions. No one, I suppose, wants to argue that animals are conscious all right, but invariably deluded about the motives of their actions. A more appropriate parallel would be with the case of a somnambulist, who, we may suppose, is apparently acting, but in fact is not conscious of anything. But this supposition, assuming it is correct, is based on clear criteria: The somnambulist is glassy-eyed and mechanical in his movements; he displays extreme shock if woken. Here, however, we have a distinction that seems to apply with equal force to animals. My—not especially bright—cat sometimes chases his tail with a degree of nonchalance that strongly suggests that he is not aware of what he is doing. For example, he makes a pass at it with his mouth and, after the tail has eluded him, stares about him in a comic state of puzzlement. He appears at least to have forgotten what he was doing; perhaps he never even knew. On the other hand one is not tempted to such a supposition when, his body quivering with intensity, he is concentrating on stalking a bird. The point of such rather banal examples is not to claim great insight into feline psychology, but merely to indicate that, provided we insist that there are criteria distinguishing conscious from non-conscious states, there is no difficulty of principle in applying them beyond the human case. It will of course be said that I am quite perversely refusing to address the real issue. When he stalks the bird is he *conscious*, or is he just acting mechanically like an 'unthinking robot'? But this is just to return to the Cartesian conception of consciousness as quite independent of any behavioural manifestations; and hence of any possible criteria; and hence, I have tried to argue, of any sense. (A distinction between conscious and non-conscious mental states entirely innocent of behavioural consequences has recently been defended by Carruthers (1989); but see Bekoff and Jamieson (1990). The specific confusions in this neo-Cartesian account cannot be addressed here.)

A somewhat more substantive concern is that animals might be incapable of awareness of anything not immediately present to them. This is an implausible suggestion about many kinds of animals. I suspect it might be less appealing if we did not suffer from such an impoverished sense of smell. If one considers an animal such as a dog without this disadvantage, it seems clear that awareness of the absent must be a major aspect of its experience. Dogs, I take it, can readily distinguish between fresh and stale scents, and can recognize both as the kinds of scents they are. Recognizing the stale scent of an opossum, say, is surely being aware of a spatio-temporally distant opossum. If, as I assume, dogs respond very differently to fresh and stale opossum scents, it would seem that it is precisely the spatio-temporal, or at least temporal, distance observed by the dog that accounts for the difference.

Of course, 'being aware of a spatio-temporally distant opossum' may reasonably be seen as no more than a bizarre circumlocution for 'being aware (or knowing) that an opossum has passed by here before'. But I take it that the reason that spatio-temporal absence has been considered important in this context derives from ideas about intentionality, that thought can be 'about' things that are not imme-diately present. And this is more obviously implied by the paraphrase than in my original formulation. (The new formulation also raises the question of the attribution of states such as *belief* to animals, which I shall consider further below.) Griffin (1984), in a similar vein, discusses some experiments on birds in which they learned to search in a variety of ways for items of food concealed by experimenters, and hypothesizes that perhaps the birds have mental images of the food they are searching for. But what is important here is the appropriateness of the thoroughly intentionalistic expression 'searching for'; not whether some—in my opinion wholly mystifying—neo-Cartesian explanation of this capacity is correct.

More mundane examples come readily to mind. I make a certain noise to communicate to my cats that I am prepared to feed them. If, as is usually the case, they are hungry, hearing this noise causes them instantly to run to the kitchen just sufficiently slowly to make sure I am following and to attempt to trip me up. It will of course be objected that this does not show that they associate this noise with (spatially distant) food. Perhaps they have just been conditioned to respond to a particular auditory stimulus with movement to the kitchen; certainly this response has been rewarded in the past. It is not altogether easy to justify the intuitions common to almost everyone who has interacted with reasonably intelligent animals other than specially bred rats and pigeons that this is a thoroughly perverse interpretation. One important ground for it is that such behaviour cannot be treated in isolation. These cats, for example, frequently exhibit the same behaviour without any stimulus, and also with other food-suggesting stimuli, such as the sound of a can-opener. Again, if they have just been fed, they may not respond at all. It strikes me that the assumption that they associate certain sounds with (absent) food, and hence, if they are hungry, go to the kitchen where food is often provided for them, is vastly more parsimonious than any attempt to reduce the phenomena to conditioned pairs of stimuli and responses. In the final section of the paper I shall offer support of a rather different kind for this sort of interpretation.

Let me turn now to a rather different, but I suspect even less problematic, issue: Are animals intelligent? I take the unproblematic answer to be, roughly: Yes, some animals have quite considerable degrees of intelligence, though no doubt some others have very little. I shall not try to offer a rigorous definition of intelligence, partly because I doubt whether such a thing is available or appropriate. A central aspect, I take it, is the ability to find solutions to problems, and to do so with some flexibility. (This is an important general theme of Griffin 1984.) By the latter, I mean to exclude responses that are invariably elicited by a particular stimulus, regardless of whether they are, in the particular circumstances, appropriate solutions to problems;

analysis of invertebrate behaviour often, though by no means always, shows highly adaptive behaviour not to be intelligent on this criterion. The criterion I have offered is, of course, an empirical one, and it might have turned out that animals were entirely stupid. The ethological literature shows this to be far from the case. The observations surveyed by Donald Griffin in his *Animal Thinking* provide an excellent source of illustrations.

There remains, nonetheless, the familiar illusion of a much deeper problem. Once again, this comes down to the idea that, however intelligent behaviour may appear to be, there is still the question whether we are dealing with a genuinely intelligent being, or merely a 'mindless robot'. As I have tried to indicate in previous contexts, I take this concern to be deeply incoherent. This misconception of intelligence has been attacked with great brilliance by Gilbert Ryle (1949). Ryle argues forcefully that intelligence is grounded primarily in intelligent *action*. The Cartesian picture, what Ryle refers to as the myth of the ghost in the machine, leads us to suppose that the intelligence of an action (which, for us, will often be a linguistic action) is not intrinsic to the action but resides in some internal and inaccessible mental antecedent of the action. Thus, for the Cartesian, an intelligent move in chess, a witty and *apropos* remark, or a smart piece of base-running are only symptoms from which we may attempt to infer intelligence. But, as we have seen in considering the so-called problem of other minds, if such an inference were needed, it would be very poorly grounded. Ryle argues, on the contrary, that intelligent performances are constitutive of intelligence. Hence, and plausibly enough, it is not an open question whether a person whose remarks and actions are consistently intelligent is in fact intelligent. To put the matter in a Wittgensteinian mode, intelligent performances are criteria of intelligence. They are defeasible criteria, in the sense that we may show that particular, apparently intelligent, performances should be attributed to luck, habit, or whatever. But it makes no sense to ask whether *all* intelligent performances might in fact fail to be intelligent.

It is in the light of these rather straightforward observations that, I want to claim, the question of animal intelligence is a simple one. There is ample evidence that animals are often capable of appropriate and flexible responses to a variety of problem-posing situations. Striking examples drawn pretty much at random from Griffin (1984) include accounts of remarkably coordinated cooperative hunting among lions (pp. 85–7), or the versatility with which captive great tits learn to solve experimentally constructed foraging problems (pp. 65–7). Such responses are not symptoms from which intelligence can, at some intellectual risk, be inferred. They are constitutive of intelligence. So, I conclude, many animals are fairly intelligent.

Turning finally to a topic about which I shall say very little, let me nevertheless say something about non-human language and the importance of its alleged non-existence. It is certain that many animals communicate to some extent with one another, perhaps almost all do to a very limited extent. It is equally certain that we have not encountered anything remotely like a human language in any other species. By the first point I mean that they convey information about such things as their emotional states (grimaces, growls, etc.) and their environments (alarm-calls and the famous waggle-dances of bees); and the behaviour by which they do this has the primary function of so conveying information. On the other hand, it is extremely improbable that any non-human terrestrial creature has any use for such things as, for example, pluperfect subjunctives, or even subordinate clauses. Exactly how wide this gulf will prove to be remains a fascinating question now undergoing investigation from various directions. Contemporary studies of social animals in the wild, and of attempts to teach non-humans fragments of quasi-human language, may throw light on the extent to which human language is a wholly novel evolutionary creation, or simply a by-product of generally highly developed mental capacities. But I shall not attempt to review these questions here. Rather, assuming that animals do not possess anything like human language (whether or not they may have the capacities for significant parts of it) I want to consider whether this lack shows that they must also be missing other central features of a mental life.

There are a variety of reasons why what I take to be quite disproportionate importance is often attached to the question of animal language. Perhaps the most important of all such reasons are, broadly speaking, political: For a variety of economic, religious, or other ideological reasons, it has been important to many people to insist on an unbridgeable gulf between humans and animals, and language has seemed the most promising instrument for achieving this. (These political aspects of the question have been particularly emphasized to me by Harriet Ritvo.) Closer to my present concerns, it has been thought (e.g. by Descartes) that language was a necessary condition of consciousness or of intelligence. Recent philosophers have suggested that language is necessary for a being to possess beliefs. (States such as beliefs, desires, intentions, hopes, fears, etc., which involve a relation between a subject and an actual or possible state of affairs, are often referred to generically as 'propositional attitudes', and treated similarly in this context.) I shall briefly consider each of these claims.

Perhaps the easiest such idea to reject is that intelligence should require language. If, as I have suggested, intelligence should be conceived as appropriate and flexible response to problem-posing situations, then it is impossible to see why this should require linguistic ability. It might be suggested that it would be impossible for an entity to show intelligence without the capacity to conceptualize the situations with which it was confronted. For the appropriateness of a response will depend on the kind of situation involved; and for the range of responses to be flexible, the entity must be able to discriminate different kinds of situations, which is, perhaps, to exhibit that it has 'concepts' of these kinds of situation. But if this is right, then the possession of concepts is a capacity independent of the ability to express them linguistically. It is true that we, as highly linguistic beings, tend to associate concepts very tightly with words we use to express them; and there are no doubt many concepts that we possess that we could not possess if we were not linguistically talented. But if we require no more of the possession of a concept than the ability to discriminate what falls under it from what does not, the connection with linguistic capacity is surely quite contingent. And the possession of 'concepts' thus tolerantly conceived is all that is required for at least a modest degree of intelligence.

Let me finally turn to the so-called propositional attitudes, belief, desire, intention, and the rest. As is conventional in contemporary philosophy, I shall focus especially on belief. The best-known recent defence of the thesis that without full-blown language a creature cannot properly be said to possess beliefs is perhaps that of Donald Davidson (1975, 1982), and I shall focus on his treatment of the question. Davidson's central argument for the dependence of belief on language is as follows. Human beliefs, at least, are deeply embedded in a very complex structure of belief. By contrast, Davidson considers the attribution to a dog of the belief that a cat is in a particular tree. (The dog in question has been chasing a cat, and is now barking up a tree, though in fact the wrong tree.) Davidson writes:

can the dog believe of an object that it is a tree? This would seem impossible unless we suppose the dog has many general beliefs about trees: that they are growing things, that they have leaves or needles, that they burn. There is no fixed list of things that someone with the concept of a tree must believe, but without many general beliefs there would be no reason to identify a belief as about a tree, much less an oak tree. (1982: 3)

It is, indeed, plausible that we would not attribute the belief in question to another human unless we supposed that he or she was in possession of the sort of general truths about trees that Davidson describes. But we should be careful about what this shows. In particular, we should carefully distinguish the question whether the dog has something like the full complement of beliefs that we would expect of another human about whom we asserted 'he believes that there is a cat up that tree' from the question whether the dog has some belief which has some significant content in common with that former belief. I think that the answer to the first question is negative, but that that is all Davidson's argument shows; and I see no reason to doubt that the answer to the second question may well be affirmative.

I shall say something more about the attribution of particular beliefs to animals in the final section of the paper, but first I should consider further the question whether we are justified in attributing any belief at all to a dog. I take this to be a question about the appropriateness of a general explanatory strategy for dealing with animal behaviour, the strategy Daniel Dennett (1971) refers to as 'the intentional stance'. Roughly speaking, this is the strategy of trying to decide what an animal is aiming to achieve and what it believes are the avenues open, and obstacles, to achieving that goal. This, of course, is the way we standardly explain the behaviour of other humans. Someone who objects to using this approach to the behaviour of other animals should, I take it, be prepared to advocate some preferable strategy. As far as I can see, the only systematic alternative would be to adopt what Dennett refers to as the 'physical stance'. That is to say, we analyse the animal as a physical structure, and determine, by appeal to knowledge of the laws of nature, how that structure will behave in response to a given set of environmental stimuli. One research programme that more or less fitted this latter model was behaviourism. The characteristics of the physical structure could be taken to have been determined by a past history of stimuli, behaviour, and rewards, and this history would determine the response of the structure to new situations. But this programme has had almost no success in understanding the behaviour of animals in anything like natural conditions, and for reasons that are well understood. Animals are more intelligent than the programme allows; they have a much more interesting internal structure—I am inclined to say, a structure of beliefs and goals—than it suggests.

There is a quite different candidate for a physical stance approach to behaviour, one that is now widely held to be very plausible, and that is to give a genuinely physical analysis of the internal structure, presumably a neurobiological account. Two points should be noted about this. First of all, no such account is anywhere near being available. So what we are considering is only a possible strategy, not a real alternative. Possibility should certainly not be discounted from a philosophical perspective, and it is certainly an important question whether this possibility is genuine. However, and this is the second point, this possibility in no apparent way distinguishes the non-human from the human situation. There are, indeed, a number of philosophers who believe that some day we will be in a position to replace our intentional stance explanations of human behaviour with physical stance explanations grounded in neurobiology. These philosophers go on to conclude that the concepts in terms of which we give intentional stance explanations—beliefs, desires, etc.—would then have turned out to be fictitious. While I am not persuaded of the coherence of this view (see Dupré 1988 and also Clark 1990), the point of present relevance is simply that no fundamental divide between the human and the non-human is implied: perhaps, strictly speaking, no one has ever really believed anything; but if so, beliefs certainly do not depend for their existence on language. And if they are simply a fiction we make do with, lacking an adequately developed neurology, there seems to be no reason for taking this fiction to be any less necessary for non-humans than for humans. I do not, of course, want to deny the obvious fact that the majority of beliefs we attribute to humans could not sensibly be attributed to non-linguistic animals. This is simply because for very many beliefs, perhaps the majority, the only possible criterion is a verbal expression. But there are nonetheless many beliefs that we attribute to both humans and non-humans on the grounds of simpler behavioural criteria; so, I want to maintain, the difference is ultimately one only of degree.

I have not attempted to exclude every possible ground for drawing a deep divide between human and animal cognition. Indeed, I have not attempted to consider the range of arguments that Davidson deploys towards the establishment of this division. So perhaps I shall conclude this section by using Davidson quite unfairly against himself. The final conclusion of the paper discussed above is that 'rationality is a social trait. Only communicators have it.' Since many animals are social, and many animals communicate, we should perhaps enlist Davidson in support of the view that there are many kinds of rational animals.

Some Hard Questions about Animal Minds

The general theme of this paper so far has been to argue that certain kinds of 'deep' mystery that appear to arise in connection with the question of animal minds are illusory. The suggestion that no non-human animal is conscious, sensate, moderately intelligent, or in possession of even the simplest beliefs can, I have tried to argue, be founded only on serious misunderstandings of what is involved in the application of mental descriptions. Thus I want to conclude that there should be no difficulty in deciding that many other kinds of animals *have* minds. However, even a more defensible interpretation of mental language can present deep and perplexing obstacles to the interpretation or characterization of non-human minds. In this final section of the paper I shall indicate a perspective on mental language that suggests that such problems, even though difficult, are at least solidly empirical and, in principle, tractable.

Since the first part of this paper depended heavily on an interpretation of Wittgenstein, it may be appropriate to introduce the present discussion with one of Wittgenstein's better-known aphorisms: 'If a lion could talk, we would not understand it' (1953: 223). This remark develops the intuition that language is deeply integrated with non-linguistic practices and behaviour. Since lions, and other animals, lead wholly different lives, their hypothetical language could make no sense to us. Does this imply—if it is true—that we are necessarily mistaken in applying terms of our language to a lion? To pursue a standard example, a possible reason for hesitation in applying the term 'pain' to a lion would be that while there is much in common between the natural expressions of pain in humans and the behaviour of lions when they are injured, there are also differences. Lions do not, I suppose, exactly cry. Moreover, again on a Wittgensteinian picture, for humans these natural expressions are often replaced by verbal ones, statements such as 'I am in pain'. Some kinds of pain attribution seem to depend almost exclusively on verbal criteria. There are perhaps no criteria for attributing a headache to a lion.

Suppose, then, that our lion found its voice and said something that we were (somehow) inclined to translate as 'I am in pain'. Why might we not be right in this translation, and thus understand the lion? One might imagine a Wittgensteinian answering that the role that such an utterance could, imaginably, play in the life of lions, and its relation to the natural leonine expressions of pain, would be different from the equivalent role of the English utterance in the life of humans. If this seems wholly implausible, it is perhaps because the behaviour associated with pain is so primitive that it really does extend to many non-human species without serious alteration. But then we should certainly be doubtful about the reference of pains for talking whales (which do not groan, still less grimace), let alone for beetles or butterflies.

There is a powerful pragmatic reason for rejecting this line of argument. Most people think it is a very bad thing to torture, or gratuitously injure, lions, whales, and perhaps even beetles, though the extent of these intuitions is notoriously variable. Presumably part of the reason for this is that we think that these animals feel pain, and pain is a bad thing. It would seem that doubts such as I have been raising about the legitimacy of applying our term 'pain' to lions or the translation as pain of some term in the vocabulary of a talking lion would threaten to undermine all such intuitions, and show that the objections to torturing animals must be incoherent. I am inclined to think that this conclusion is a *reductio ad absurdum* of the line of argument that purports to demonstrate it.

But it is also fairly clear that the threatened conclusion cannot be a legitimate inference from the arguments under consideration. Wittgenstein's argument does not show that humans only come to feel pains at the point when they learn to talk about them; on the contrary, they show that if pain did not pre-exist pain language, there could not be such language. Hence whatever we make of the roars and so on of injured lions, their status as expressions of sensation, at least, cannot be undermined by considerations concerning the linguistic incompetence of lions.

I want to suggest that taking the Wittgensteinian perspective a stage further points to an attractive resolution of this difficulty. It is extremely difficult to get rid of the intuition that what we are really

concerned with in attributing pain to the lion is a correct characterization of what is going on in the lion's mind: what, if it could only
talk, and if we could only understand it, the lion would refer to by
its word 'pain'. Put in another contemporary idiom, we are trying to
speculate about part of 'what it is like to be a lion'; and that, as has
been discussed by Thomas Nagel (1974), is a hard thing to do. But
I have recommended rejecting all these interpretations of the problem, and I suggest that we focus instead on the question, What do
we mean—if anything—by attributing pain to a lion? We should
remember, in other words, that 'pain', even when applied to lions, is
a word in *our* language. And if *this* is the question, then it can be
seen that our difficulty in understanding the dicta of hypothetical
talking lions is completely irrelevant.

Of course, serious doubts might be raised about whether our
term 'pain' *should* be extended to lions, whales, or whatever. The facts
alluded to above, that the criteria for such applications will differ in
some important respects from those appropriate for humans, show that
this is an extension of our concept, not a paradigmatic use. But I think
that from this perspective it is clearly a natural and obvious extension, comparable, perhaps, to our extension of the term 'conversation'
to telephone conversations and even rapid exchanges of computer
messages. Lions and other animals avoid things that cause them pain,
withdraw rapidly from painful stimuli, and generally respond to
pains in ways that are more or less analogous to human responses.
But perhaps most important, to pick up an earlier point, the concept of 'pain' fits into broader aspects of our conceptual scheme, most
especially the ethical. We think that causing pain is a very bad thing,
because it is a sensation that sentient beings greatly dislike. Since lions
are clearly sentient, and show every sign of disliking the experiences
which, I am suggesting, we refer to as 'pain', we should avoid causing them these experiences. This is another very powerful reason for
including their relevant experiences under our concept of 'pain'.

I think this perspective, if it is accepted, should also defuse some
of our worries about the attribution of cognitive states to animals.
When we say, to return to an earlier example, that the dog believes
that the cat went up the maple tree, we are, obviously enough, saying
something in English. Whether such a statement is appropriate or not
must surely depend, then, on whether the dog satisfies criteria that
we would usually employ in attributing such a belief to a human. It
does not depend, for example, on whether the dog is entertaining some

proposition in Caninese that would be correctly translated as: 'The cat went up the maple tree'. Indeed, it seems pretty obvious that we might well attribute this belief to another human on exactly the kinds of grounds proposed for the dog. (We see a man chasing a cat, shouting abuse, and swinging at it with a stick; the cat darts up an oak tree, and the man continues yelling up the maple tree and shaking his fist, etc.) Of course the majority of beliefs that we attribute to humans probably do depend essentially on linguistic criteria; but many do not, and such may very well be attributed to non-humans.

There is a slightly different kind of worry that might still be raised, that suggests a serious and pervasive risk of error in such attributions. Richard Jeffrey (1985), in discussing Davidson's example, suggests that although the dog may not have the concept of a tree that Davidson describes it may have a concept of its own that the maple tree falls under; he suggests 'marker that a scratcher can disappear up'. Although, as I have been arguing, I do not think the question at issue is illuminatingly construed as one about the phenomenological states of dogs or lions, this suggestion does point to a real difficulty. This is simply that it is easy to be wrong in giving intentional explanations of the behaviour of non-humans. Here I mean to identify the point at which armchair theorizing ends, and the difficult empirical task of cognitive ethology begins. To be successful, as opposed to merely logically intelligible, in attributing beliefs to other animals, we need to know a great deal about the animals we are talking about. If we do not know what their perceptual capacities are, and what features of the environment they are capable of discriminating (see Rosenzweig 1990); the goals that such animals often pursue; the level of intelligence they are capable of bringing to bear on the pursuit of these goals; their tendencies to stereotyped or habitual responses to certain kinds of situations; and so on, we are likely simply to be wrong in our suggestions as to what their cognitive states are. Hence it is unsurprising that we can make such attributions much more easily and widely in the case of our conspecifics, and after that we are inclined to feel more comfortable with the beliefs of cats and dogs—about which we feel we know a fair bit—than with whales or bats. But I see no reason why, as we come to know more about other creatures, we should not come to be very successful in giving intentional stance explanations of their behaviour—explanations in our own language, which is just as well since they probably have none, and if they did we could not, perhaps, understand them.

11

Conversations with Apes: Reflections on the Scientific Study of Language

Contemporary attempts to teach a form of language to great apes date from the work of Beatrice and Allen Gardner. While a number of earlier attempts were made to teach spoken languages to domestically raised chimpanzees,[1] these were uniformly unsuccessful. The insight pursued by the Gardners was that the failures might be due as much to the apes' lack of suitable vocal mechanisms as to their lack of linguistic aptitude (Gardner and Gardner 1971). They therefore attempted instead to teach their charges—notably the chimpanzee Washoe, undoubtedly the best-known of the talking apes—American Sign Language, or Ameslan. The Gardners claimed considerable success for this project, Washoe acquiring a vocabulary of well over a hundred signs[2] and significant conversational skills. Since then a number of other chimpanzees have been instructed in sign language (Fouts 1973: 978–80; Terrace 1987) as well as a pair of gorillas (Patterson and Linden 1981) and an orang-utan (Miles 1983). A rather different strategy has been pursued by David Premack (1972), Duane Rumbaugh (1977), and Sue Savage-Rumbaugh (1986), who attempted to teach chimpanzees wholly artificial symbolic systems. In the first case this involved learning to place variously coloured and shaped metal pieces in sequence on a magnetic board, and in the latter two, pressing keys on a specially designed computer keyboard. The difference between the sign language and other approaches will be discussed further below.

I am grateful to Regenia Gagnier and Debra Satz for pointing out a number of obscurities in an earlier draft.

[1] For a brief summary, see A. Premack (1976, ch. 2).

[2] The actual number may be much higher, perhaps several hundred. Criteria of various degrees of stringency are applied to decide whether the animal has genuinely mastered particular signs.

Numerous interesting questions are raised by this research. I shall divide the present discussion into three broad categories. First, I shall briefly consider what it is these various apes have really learned to do. Second, in the main part of the chapter, I shall consider a variety of objections that have been posed to claims that such apes have acquired genuinely linguistic abilities. These objections reveal how the research under consideration raises important methodological questions about the scientific study of language. And finally I shall consider some of the varied goals and interests underlying this research programme and its criticisms.

What do Apes Learn?

The assertion that apes have, to some limited degree, mastered a language has engendered enormous controversy. One major issue has concerned whether apes ever acquire any kind of syntactic capacity.[3] Given the widely held Chomskyan idea that the indefinitely productive powers of language, grounded in syntax, are of its essence, this has often been interpreted as showing that pongid utterances[4] are not really linguistic at all. On the other hand, it seems widely conceded within the debate that the question what is required for syntactic competence is an extremely obscure one, and I shall here sidestep it.

What is fairly clear is that apes can be taught to use quite substantial repertoires of symbols (the gorilla Koko, apparently the star in this area, is said to have a vocabulary of around 150–600 words, depending on the strictness of the criteria employed; Patterson and Linden 1981: 84). Slightly more contentiously, it may be said that apes can perform certain speech-acts. These two claims are sufficiently illustrated by the kind of performance of which these apes are most widely agreed to be capable, that of demanding. Requesting various foods and drinks, tickles, and other preferred amusements figure largely in the reported utterances of language-trained apes. Given that these requests are effected by the use of essentially arbitrary symbols (signs, selection of keys marked with geometric symbols, etc.), it seems clear that these apes can use symbols. Against this it is sometimes argued

---

[3] Notable among those questioning the syntactic capacities of apes has been one of the leading researchers in the area, Herbert Terrace (1987).

[4] I use the term 'utterance' throughout to refer to the alleged linguistic productions of apes. Obviously apes do not, strictly, 'utter', but this usage has become standard.

that all that has happened is that the apes have been conditioned, in a crudely Skinnerian way, to bring about certain desired results. On this view an ape said to be signing 'Give me banana' is doing essentially the same as a rat trained to press a red lever to obtain a food pellet.

One might perhaps retort that the rat has indeed been taught that red means 'pellets', and thus has acquired a minimal semantic competence. Generally the intended force of the criticism is that the ape, and a fortiori the rat, does not know that its sign, say 'X', means banana. And certainly one is not inclined to attribute this knowledge to the rat, presumably because it seems more parsimonious to attribute to the rat merely the causal belief that pressing the lever produces food. The critic wants to say the same of the ape, that signing 'give me banana' is essentially similar to shaking the tree to make the bananas fall off. Presumably rebuttal of this criticism should not be taken to require showing that apes have explicitly semantic beliefs, e.g. the belief that 'X' means bananas. Mastery of this kind of semantic ascent would surely exclude most small children and many adults from our linguistic communities; sentences such as ' "banana" means bananas' are pretty sophisticated bits of philosophical gadgetry. A much more reasonable demand would be that to qualify as using 'X' as a symbol one should be able to do more with it than merely make requests (though compare the notorious language games in Wittgenstein's *Philosophical Investigations*, §§1–7).

It would appear that this latter demand has been quite thoroughly addressed in recent ape-language research. The fact that an ape's ability to use a symbol in a particular way does not entail its possession of all the capacities we might naturally associate with knowing the meaning of a common noun is nicely demonstrated by some of the experiments of Savage-Rumbaugh (1986). For example, apes can be taught to produce utterances appropriate for requesting different available kinds of food, while giving no indication that they can respond appropriately as addressees of the same utterances, e.g. in selecting the named object from a group of objects. (A similar separation between linguistic production and reception has also been noted in the linguistic development of children.) On the other hand, this is equally clearly not the typical case for trained apes. In the experiments referred to above, once the animals acquired the capacity of responding to, as well as producing, particular kinds of utterance, they had no trouble in generalizing this facility to newly acquired signs.

Much more striking illustrations of the flexible use of signs can be drawn from the reports of Patterson's work. She reports cases of descriptions by her gorillas of events present and past, and the production of jokes, threats, and insults. Koko is reported as using signs while playing with her dolls (though apparently she is embarrassed to be discovered doing this), and even as discussing death:

MAUREEN (an instructor). Where (do) gorillas go when (they) die?
KOKO. Comfortable hole bye.
MAUREEN. When (do) gorillas die?
KOKO. Trouble old.

(Words in parentheses do not actually appear as signs in the original Ameslan.) (Patterson and Linden 1981: 191)

Without constituting brilliant dialogue or philosophy, such reported interchanges surely suggest more than conditioned reflexes. However, their significance has been seriously questioned. I shall now look in more detail at the kinds of attacks that have been launched against the conclusions that these apes reveal any genuine linguistic capacity.

Criticisms of Ape-Language Research

It will be helpful in considering the criticisms of research on the linguistic capacities of apes to distinguish sharply between the sign-language projects of the Gardners, Fouts, Terrace, and Patterson, and the artificial languages of Premack, Rumbaugh, and Savage-Rumbaugh. Somewhat crudely, one may see the former as motivated by the goal of maximizing the level of communication with the subjects, as against the latter, in which greatest importance was attached to the possibility of acquiring clean, unambiguous, and well-controlled data. Thus in the first category, taking as exemplary the work of Patterson, one is struck by the establishment of rapport between subject and researcher over a long period of time. Patterson emphasizes, for example, ways in which the communicative intent of her subjects can be revealed, at least to the experienced observer, in ways that transcend the 'literal' interpretation of strings of signs; novel and unexpected productions are given particular emphasis. At the other extreme, in Premack's experiments the possible behaviour of the subjects is tightly constrained by the experimenter. The ape is typically given access to only a very limited number of signs, thus facilitating quantitatively tractable predictions and analysis of the ape's response to particular promptings. This sharp dichotomy is something of a caricature: on the one hand, the artificial-language researchers discuss the affective connections established with their charges, and Savage-Rumbaugh, for instance, attaches considerable importance to unanticipated and spontaneous productions (the chimpanzees in her experiments appear to have had continuous access to a larger range of signs than those in Premack's, though not the full repertoire available to Koko or Washoe); on the other hand, Patterson, the Gardners, and other sign-language experimenters devote serious effort to fairly rigidly controlled tests of their apes' vocabulary. However, I think the contrast fairly indicates the emphasis of the different experiments, and also serves to differentiate the main lines of criticism of these experiments.

Two prominent critics of ape-language research, Jean Umiker-Sebeok and Thomas Sebeok, usefully distinguish three major lines of such criticism (Umiker-Sebeok and Sebeok 1980: 9). These are (1) inaccurate observations and/or recording of ape behaviour; (2) the over-interpretation of ape behaviour; and (3) the unintended modification of an animal's behaviour towards desired results. I shall look next at each of these kinds of criticism.

Complaints about inaccurate observation and reporting are directed most especially against the research using sign language. David Premack (1986: 32), for example, cites a study by M. S. Seidenberg and L. A. Petitto (1979) (collaborators of Terrace) suggesting that comparisons between tapes of Washoe's performance and the published reports of the Gardners often fail significantly to agree. For example, where Washoe is reported as replying to a question with the signs for 'you me', the actual response as seen on the videotape was 'you me you out me'. Premack remarks that 'the Gardners appear to have assumed . . . something like "if we can extract from the ape's garbled message what the ape is trying to say, so can the ape"'. But while this assumption may certainly sound very 'unscientific', further reflection may lead one to wonder. After all, if what the ape is producing really is a kind of language, then we should surely not be surprised that the listener will typically be required to contribute a measure of interpretation to the communicative interaction. Literal transcriptions of conversations even between linguistically competent human adults typically look very different from grammatically correct written language; yet this does not lead us to doubt that successful and intentional communication is taking place. Even with the significantly more aberrant utterances of human children, we are prepared to believe that parents often understand what is being said.

Two particular circumstances reinforce the reasonableness of charity in these cases. First, it is often remarked that apes, especially chimpanzees, are typically operating at a high level of activity and even excitement. The repetitive linguistic style which often *is* literally transcribed seems highly consistent with this trait.[5] More significantly, the particular characteristics of a signed language such as Ameslan seem to necessitate greater subtleties of speaker interpretation even than normal English conversation.[6] Native users of sign, it appears, use a number of cues such as location of the sign in the space around the body, direction of glance, or facial expression to convey various syntactic and other aspects of an utterance that cannot economically be conveyed with independent symbolic units. Thus the translation from sign to a spoken language is intrinsically more complex even than that between spoken languages. So the claims that those who have worked with particular signing apes for long periods of time can acquire considerable and subtle capacities to interpret what they are saying cannot simply be dismissed on the basis of naively literal interpretation.

The preceding points begin to illustrate the difficulties with the second general line of criticism, that the behaviour of apes is often wishfully over-interpreted in the light of the experimenter's expectations. They do not, of course, show that any such interpretations of ape utterances in sign language are in fact well grounded. Indeed, it would be beyond my competence to make any general judgement on this question. What I do want to emphasize at this point is that there appear to be fundamental conflicts between intrinsic features of this kind of research and commonly held ideals of scientific enquiry. Most obviously, it is widely supposed that data requiring interpretation, even debatable interpretation, by the researcher are scientifically unacceptable. Of course, since the decline of classical positivism—

[5] Striking confirmation of this construal is provided by the work of Miles (1983) on the orang-utan, a much more languid and phlegmatic beast. Her subject, Chantek, rarely made immediate repetitions. In a very similar vein, it is sometimes suggested that chimpanzees cannot really converse, partly on the grounds that they so frequently interrupt their human interlocutors. Chantek, apparently, had no more tendency to interrupt his trainer than vice versa.

[6] See e.g. Hill (1980: 336); Terrace (1987: 237–8). Terrace provides a useful general description of sign language (1987: 235–54).

indeed largely responsible for that decline—there has been a wide-spread realization that to some extent all data are interpreted in the light of some theoretical background. Observations and descriptions of electrons, tectonic plates, or pecking orders cannot exist independently of the theoretical contexts in which these terms are given significance. Nevertheless, various factors commonly maintain the trappings of objectivity despite this realization. First, the theoretical background may, at a particular historical moment, be uncontroversial. Thus, for example, contemporary anti-realism in philosophy of science notwithstanding, the observation of an electron passing through a cloud chamber is often considered paradigmatically objective. Second, it is considered very important that scientific data be replicable. Without in any way mitigating the role of interpretation in the description of data, this does provide a strong kind of intersubjectivity. Any researcher, it is supposed, can confirm that a certain kind of experiment does indeed have a certain kind of outcome.

It is clear that neither of these bases for claims of objectivity can readily be applied to ape-language research. What the controversial interpretations of ape signings require is the theoretical background assumption that the apes, when signing, are attempting to communicate something. But far from being uncontroversial, this is precisely what the critics are inclined to deny. Yet clearly it would be impossible to investigate the possibility that apes are saying something without at least the working hypothesis that this is what they are trying to do. Replicability is an issue that I shall discuss further below. But it is worth pointing out here that it has a quite paradoxical relation to this kind of study. Roughly, the more impressive a bit of linguistic behaviour is, the less likely it is to be replicable. The exact contrast which underlies scepticism about ape language is that between language as a creative and spontaneous form of behaviour, and stereotyped performances which may be taken as no more than causal and semantically innocent manipulation of the environment. By definition the latter, but not the former, would be susceptible of predictable and reliable replication.

In addition to these apparently impassable barriers to certain conceptions of acceptably objective data generation, there is a feature of this research that may well seem more positively, and objectionably, subjective. Ideals of scientific objectivity typically include a central role for the detached and dispassionate observer. Yet it is clear, and generally admitted, that someone who has devoted a substantial portion of his or her life to working with a highly intelligent and appealing creature will be anything but detached and dispassionate (see Umiker-Sebeok and Sebeok 1980: 5–8). An obvious feature of this affective connection between researcher and subject is that the researcher will typically be anxious for the success of their subjects in language acquisition—very likely as anxious as typical human parents that their children be educationally successful. But an emotional commitment to a particular outcome of an experiment is complete anathema to conceptions of scientific objectivity. (Whether scientists are typically wholly disinterested in the results of their experiments is hardly beyond question. Given the structure of rewards in the scientific professions, this would constitute almost incredible high-mindedness. But that is another topic.)

But as with the other aspects of objectivity one may suspect that these apparent drawbacks are inevitable in this research. It is after all possible that the ape, like the human child, will learn only if someone with whom it feels an affective connection shows some sign of caring that it does so. Much of human learning might well be inaccessible to a thoroughly disinterested and 'objective' study. In addition, to return to an earlier point, it might well be that the possibilities for comprehension of the linguistic efforts of an ape may be vastly greater for someone intimately familiar with its interests and its idiosyncrasies than for the dispassionate and disinterested scientific observer.[7]

The general tenor of the preceding discussion may be summarized as follows. It is highly plausible that what these criticisms of ape-language research really illustrate are very basic conflicts between ideals of scientific research and certain kinds of language study. Everyday language is, after all, typically learned in a highly emotive context, and affective aspects of communication, even between competent speakers, can hardly be divorced totally from the aseptically semantic. Moreover, language without interpretation is plainly an incoherent conception; and this may be so in a sense that precludes the study of controversial candidates for language in ways that meet scientific conceptions of freedom from interpretation. Perhaps we might have more idea of the linguistic capacity of apes if the research had been carried out by literary critics.

---

[7] Miles (1983: 57) and others have suggested that the failure to recognize the importance of establishing rapport with his subject, Nim, may account for some of Terrace's negative results.

Research using artificial languages has been motivated, to a considerable extent, by the attempt to avoid these divergences between the methodology of the Ameslan research and assumed norms of scientific enquiry. Sequences of coloured shapes or computer keystrokes provide, by contrast, thoroughly 'clean' and unambiguous data. Another virtue of the approach is apparent from the preceding discussion. I noted above that a prerequisite for interpretation of the more interesting and suggestive utterances of apes was the working hypothesis that the apes were trying to communicate. This suggests an obvious circularity, though not necessarily a vicious one, in the process of interpretation. It is an attractive goal to ground this hypothesis in some more rigorously analytic demonstration that apes can, indeed, use symbols with communicative intent. Prima facie, just such support can indeed be gained from the artificial-languages research.

Probably the most analytically detailed study of ape symbol acquisition is the work of Sue Savage-Rumbaugh (1986). The most distinctive feature of this research is the attempt to break down and test independently a range of different possible ways in which an ape may be using symbols. A noteworthy result of this work is the observation that apes often cannot, without specific training, generalize from one kind of symbol use to others, which leads Savage-Rumbaugh to be sceptical whether all the apes studied in the experiments under consideration have acquired this full range of skills. However, her work does appear to show that apes can learn to use and understand symbols, refer to objects that are not present, make spontaneous comments, and announce their intended actions. In one particularly interesting series of experiments, an ape was required to request from another tools (keys, straws, magnets, etc.) which it required to gain access to food, demonstrating its ability to use language to facilitate cooperative enterprises. And apes have even been able to teach symbols to one another. (Patterson and Linden amusingly recount an attempt by Michael, a later gorilla addition to her experiments, to instruct one of his trainers in the use of a sign (Patterson and Linden 1981: 170). Fouts reports attempts by Washoe to teach sign to an adopted infant chimpanzee (Fouts 1983: 1–3).) Finally, it appears that some apes acquire a significant understanding of spoken English. Patterson and Linden also claim this ability for Koko, though Savage-Rumbaugh, on the basis of negative results with her principal subjects (the chimpanzees Sherman and Austin), views this claim with scepticism. However, Savage-Rumbaugh does report more recent work with the pygmy chimpanzee (*Pan paniscus*) Kanzi, showing that this ape has acquired an understanding of a number of spoken words (Savage-Rumbaugh 1986: 382–97). (This rather remarkable animal apparently acquired significant linguistic ability, including this partial comprehension of spoken English, with no deliberate training. Savage-Rumbaugh speculates that this rare and little-studied species may have substantially greater linguistic aptitude than the common chimpanzee (*Pan troglodytes*). Patterson's occasional suggestions that the gorilla is more talented than the common chimpanzee cannot be dismissed out of hand. And orang-utans, despite their more languid temperaments, are reported to do better than either chimpanzees or gorillas on various cognitive tests (Miles 1983: 47). Species partisanship is an amusing occasional subtext to this topic.)

Savage-Rumbaugh's research is richly adorned with all the trappings of respectable scientific research: controls of various kinds, 'double-blind' experiments, careful statistical analyses of data, and unambiguous recording of data by computers. The evidence seems impressively marshalled that apes can, if properly instructed, communicate, and do so with full intention.

However, these scientific virtues have not sufficed to silence the critics. This brings me to the third, and perhaps most pervasive, kind of criticism of ape-language studies, the problem of inadvertent cueing, or manipulation of the animal to produce the desired result. This is sometimes referred to as the 'Clever Hans effect' after the notorious performing horse that convinced many that it was capable of doing arithmetic. Hans, when given an arithmetical problem, would tap his hoof a number of times corresponding to the solution of the problem. Investigation eventually revealed that Hans responded to extremely subtle cues from his interrogators that he had tapped long enough, and thus learned to stop at the correct point. More generally, it is known that animal trainers can develop extremely subtle ways of influencing and controlling the behaviour of their charges. Although great efforts have been made by ape-language researchers to eliminate these possibilities, it seems very difficult to devise experiments that eliminate every possible channel of communication other than that intended by the researchers. In almost all such experiments a researcher is present with the ape. Indeed, this is hardly avoidable. Umiker-Sebeok and Sebeok, noting that apes do not normally sit quietly through their tests, remark:

Experimenters must spend a good deal of time interacting with the animal just to get it under sufficient control to enable them to administer the test, hardly what one would call ideal experimental conditions. If cueing is feasible even when a subject is sitting still and attentive, it is even more so under the chaotic circumstances created by an ape's natural response to such man-made rules. (Umiker-Sebeok and Sebeok 1980: 44 n.)

Though it may be deplorable that apes show so little dedication to the advancement of science, reluctance to sit quietly through batteries of psychological tests is hardly indicative of a lack of intelligence.

Researchers often attempt to eliminate the possibility of cueing by using so-called double-blind strategies, in which the observer of the ape's performance is distinct from the experimenter who poses the problem, and thus does not know the correct response. This strategy can produce problems of its own, most notably that the elaborate and artificial procedures involved are likely to discourage the ape's cooperation. Patterson reports that Koko frequently did refuse to cooperate in such tests and would, for example, on occasion give the same answer to every question. This provides a striking instance of the problem of the dependence of interpretation on antecedent belief. To Patterson, perfectly convinced that Koko is capable of performing the task (identifying familiar objects), it is obvious that Koko is expressing her resentment of a boring activity. Umiker-Sebeok and Sebeok, on the other hand, for whom Koko's competence is extremely questionable, wonder sceptically whether these sessions were included as sets of false responses in the analysis of the experiment.

Even when the apes do cooperate, the sceptic is likely to be unconvinced. The more complex the experimental situation becomes, it almost seems, the more possible channels for unintended communication are opened up. It is not hard, for example, to imagine cues the supposedly blind observer may use to guess what the ape 'means' (perhaps its non-linguistic behaviour). And indeed, as Umiker-Sebeok and Sebeok emphasize, the experimenters, outside observers, blind observers, naive observers, etc. do not necessarily exhaust the *dramatis personae*. In the interests of producing hard and objective data (or press copy) there is usually a camera or a video recorder being operated; no doubt the operator will often understand the experiment well enough to know what response is expected, and to provide a possible source of cues to the animal.

Central to the proliferation of these doubts is an important methodological issue. This is the commitment to explanatory parsimony

(Umiker-Sebeok and Sebeok 1980: 14–21). Occam's razor is alleged to cut away at the claims of the ape researchers in two directions. First, following on the preceding discussion, if there is some channel through which a human observer or participant might have led the ape to the correct answer to a question or an appropriate utterance, then it is assumed more parsimonious to conclude that this happened than that the ape has exhibited the ability to produce its utterance or response unaided. Second, Occam's razor is commonly employed against the more creative and innovative uses of language by apes, which, it is suggested, may more parsimoniously be attributed to error or luck. A widely cited example is the production by Washoe of the signs 'water bird' when first confronted with a swan. Critics note that Washoe was presented, on this occasion, with both water and a bird, so it is gratuitous to suppose that her utterance involved the imaginative synthesis of referring to the swan as a water-bird. Rather differently, Patterson reports that Koko, in a particularly obstinate mood, was refusing to produce the sign for drink (thumb against the mouth with fist clenched), which she had previously made thousands of times. Eventually, grinning, she made the sign, but to her ear rather than her mouth. Patterson interpreted this as an exercise in humour (Patterson and Linden 1981: 77). Predictably, critics (e.g. Umiker-Sebeok and Sebeok 1980: 16) see this as more plausibly interpreted as a mistake.[8]

---

[8] On the topic of pongid recalcitrance, Patterson observes (Patterson and Linden 1981: 6) that 'Koko has often been driven to her most creative uses of language through her obstinate refusal to submit to dull routine.' In illustration, she cites the following dialogue. Cathy, a trainer, had signed to Koko, 'What's this?', pointing to a picture of Koko.

> *Gorilla*, signed Koko.
> *Who gorilla?*, signed Cathy . . .
> *Bird*, responded Koko.
> *You bird?*, asked Cathy . . .
> *You*, countered Koko, who by this age was frequently using the word *bird* as an insult.
> *Not me, you bird*, retorted Cathy.
> *Me gorilla*, Koko answered.
> *Who bird?*, asked Cathy.
> *You nut*, replied Koko, resorting to another of her favorite insults. (Koko switches *bird* and *nut* from descriptive to pejorative terms by changing the position in which the sign is made from the front to the side of her face.)
> After a little more name-calling Koko gave up the battle, signed *Darn me good*, and walked away signing *Bad*.

It is easy to see how this little anecdote will be seen as failing a range of desiderata of scientific method. My general point, however, is that this is insufficient to show that it is impermissible to take such reports at face value.

It is hard to adjudicate particular disputes. But one point should be stressed. Parsimony is hardly an objective, theory-independent concept. *Why* is it more parsimonious to assume some complex and covert channel of communication in a double-blind experiment than to suppose that the ape knows what it is doing? If you believe that the ape is in fact capable of the performance in question, the latter explanation is surely more parsimonious. Again, in the Koko anecdote, to Patterson, confident that Koko could sign 'drink' if she wanted, the interpretation as a joke is quite natural. The sceptic, disinclined to credit the linguistic aptitude of the ape, will tend to view any alternative explanation as more plausible. Either way, what interpretation is natural, or 'parsimonious', depends heavily on antecedent belief. It will, of course, be objected that the charitable assumption is question-begging, since that is what it is the object of the experiment to reveal. But this is irrelevant to the prior probabilities that one may attach to the apes having, or lacking, the sort of capacity being investigated.

I should perhaps say that there is more of interest in the Clever Hans phenomenon than my discussion may have suggested. The general point is that, independent of controversial questions about linguistic apes, there is a great deal of communication possible between humans and animals. Sebeok remarks that 'Two way zoosemiotic communication is thus not the issue, but such communication by *verbal* means between man and animalkind is another matter' (Sebeok 1980: 426). Sebeok's objections to the ape-language research are grounded in his view that 'the Clever Hans effect informs, in fact insidiously infects, all dyadic interactions whatever, whether interpersonal, or between man and animal, and by no means excepting the interactions of living organisms with a computer' (Sebeok 1980: 426). (One wonders only why this should be 'insidious'; what linguistic essence should ideally be distilled from this non-linguistic noise?) He also stresses that the range of non-verbal means of such communication is yet poorly understood. But while it is possible that many ape-language researchers may be naive about these means of communication, it is difficult to see why their existence should cast doubt on the kinds of artificially established channels of communication they assert. Indeed, one would think the opposite. Thus I do not see how the Clever Hans effect, for all its intrinsic interest, need push us towards a sceptical interpretation of purported ape utterances.

One final point about the Clever Hans objections is particularly important. As with Clever Hans himself, the kind of performance for which cueing is most plausible is that in which decisively correct or incorrect answers are available. Thus, perhaps ironically, it is the experiments which offer clean, unambiguous data that seem most vulnerable to this line of attack. By contrast, when the ape produces an utterance that is novel and unexpected, this kind of criticism seems wholly inapplicable. (Thus in these cases the critic moves to the very different ground of suggesting chance or error.) Patterson's report of Koko chatting with her dolls, but rapidly ceasing on discovering that she was observed, is hardly to be understood as unconscious cueing by the experimenter. This will, on the other hand, be dismissed as 'anecdotal', the most damning term in the lexicon of scientific norms, applied to reports which most signally fail to meet desiderata of adequate controls and replicability (see, for example, the dialogue cited in n. 8). The irony is that it is precisely the predictable, replicable responses that *do* meet these desiderata that are vulnerable to the suspicion of cueing. It would seem, then, that ape-language research is impaled squarely on the horns of a methodological dilemma. On the one hand, the more controlled and predictable the behaviour of the animal is made, the harder it becomes to fend off the accusation of manipulation, conscious or otherwise. On the other hand, the more freedom the animal is given, and the more spontaneous and uncontrolled its utterances, the further the reports of its behaviour will sink into the scientific netherworld of the 'anecdotal'. A more optimistic interpretation is possible, however, that these kinds of research are mutually supporting of the conclusion that the ape is, indeed, intentionally and often successfully engaging in modest feats of linguistic communication. To the extent that methodological constraints threaten to preclude a priori the possibility of establishing this result, we should perhaps rather question the methodology.

I believe that there are, indeed, important grounds for deep suspicion of the appropriateness to this kind of research of various of the methodological norms that have been mentioned in the preceding discussion. Tying together ideas of objectivity, replicable versus anecdotal results, and the undesirability of affective or 'interpretative' relations between the experimenter and the subject is a conception of the appropriate role of the scientific investigator nicely expressed by H. Hediger, another prominent critic of ape-language research. Hediger writes: 'The ideal condition for all such experiments would be complete isolation of the experimental animal from the leader of the experiment' (Hediger 1980: 409). But if, as is surely not unreason-able, the goal of the 'leader of the experiment' is to communicate with the experimental animal, this seems a serious restriction. If one supposes further that linguistic communication is inextricably involved with a range of expression beyond mere stringing together of signs—as, I assume, do Umiker-Sebeok and Sebeok, though such expression will no doubt fall within the proscribed category of 'cueing'—the demand is, in addition to being impossible, misguided in its motivation. This difficulty can, as I have suggested, be seen in relation to a very general picture of the role of the scientist. The ideals of scientific enquiry I have been discussing constantly presup-pose the picture of an active (though impartial) observer set against a passive object of study. But in fact, most obviously though not only in the case of sign-language research, what is being investigated in ape-language studies is *interaction* between two intelligent subjects. Perhaps it will be thought that the very question at issue is whether the ape *is* an intelligent subject, or rather a passive object, respond-ing mechanically if complexly to its stream of inputs. I shall say some-thing bearing on this issue in the conclusion. It is my impression, though, that researchers involved in the ape-language experiments would not, with good reason, take this question very seriously.

The Goals of Ape-Language Researchers and their Critics

As interesting as the methodological prejudices revealed in the debate over ape-language research is the range of objectives detectable in both the practitioners of this research and their critics. Some of these motives and concerns, laudable or otherwise, are tangential to my pre-sent interests. Laudably enough, for example, Savage-Rumbaugh's work is directed towards improved methods for teaching language to mentally impaired children. This kind of goal is quite frequently an explicit component of ape-language research. Perhaps less laudably, one can detect some fairly overt disciplinary infighting. Umiker-Sebeok and Sebeok, for example, discussing some research on dolphins in a context which leaves little doubt that they intend the same to apply to ape research, remark that by contrast with these unpromising efforts, computers will 'within the next decade' be able to talk in, if not listen to, English. They continue: 'for computers to achieve the implied level of sophistication, vast but justifiable funding will be required;

money spent on chimerical experimentation with speechless creatures of the deep to be hominified is, however, tantamount to squandering scarce resources' (Umiker-Sebeok and Sebeok 1980: 27). As they immediately point out, apes, too, are expensive to maintain; economic competition in the multi-billion-dollar world of contemporary science is, perhaps unsurprisingly, often not far beneath the surface of scientific controversy.

My present interest, however, is rather with the questions about the natures of apes and humans which dominate the official rationales for this research. The subsequent discussion can naturally be divided between the questions, What can this research tell us about apes? and, What can it tell us about humans?

I am inclined to some scepticism about how productive a method of learning about apes the research under consideration is likely to prove. As Umiker-Sebeok and Sebeok remark in the quote above, the process to which thes apes are being subjected is one of domestication or even 'hominification'. It is not quite correct to refer to these animals as 'domestic' apes, since domestication is best seen as the coevolution over a substantial period of time of a symbiotic relationship between humans and another species; merely 'socialized' animals are a rather different matter.[9] But neither domestic nor socialized animals are a reliable model for learning about the natures of their free-living relatives without careful attention to the effects of these processes on their behaviour. Domestic cats and dogs may, by now, be a sufficiently interesting quasi-natural kind to deserve study in their own right, and their relation to undomesticated relatives can be systematically explored; none of this can plausibly be said for so rare and exotic an artefact as the language-trained ape. In the case of apes we would surely do better to consider the observations of primatologists such as Jane van Lawick-Goodall or Dian Fossey, who have devoted years of their lives to the study of these creatures in their natural habitats. It is true that despite this extensive observation, the means by which apes in the wild communicate remain shrouded in obscurity. There is little doubt that there is much here to be learned. A quite elaborate system of calls used by vervet monkeys, assumed to be much simpler animals than the great apes, to identify differentially various kinds of predators has been described in some detail (see Seyfarth

---

[9] For a detailed discussion, see Daniels and Bekoff (1990). I am grateful to Marc Bekoff for drawing my attention to this distinction.

1984; Cheney 1984), and is only one of a wide range of animal communication systems that are at least partly understood. One further
point that well illustrates the communicative capacities of untrained
primates is the following. An important criterion of intentional
communication, as has often been noted (see e.g. Wittgenstein 1953,
§§249–50), is the possibility of communicating deceptively. Duane
Quiatt (1984) has usefully surveyed an impressive range of evidence
that apes, and even monkeys, are quite up to this feat. It would be
remarkable, then, if social animals as intelligent as chimpanzees or
gorillas did not employ some quite sophisticated means of communication. But apart from providing minor encouragement to the
search for these means, there is no reason to expect the linguistic
achievements of highly socialized apes to throw much light on this
question. The fact that the odd chimpanzee has been propelled into
space will hardly help us understand how chimpanzees manage to
swing through the trees.

Much deeper philosophical interests, however, are frequently, if
not universally, connected with interpretations, positive or negative,
of ape-language experiments. Frequently the name of Descartes is
invoked to connect these experiments with long-standing and continuing debates about the internal lives of animals. One pervasive
theme is that the research will illuminate the supposed problem of
the internal lives of animals, or their lack of any. Of most obvious
pertinence to the present topic is the Cartesian thesis that language
is the only reliable indicator of genuine mental process. Also relevant
is the corollary, that behaviour in general has only contingent relations
to the underlying mental processes that it might be held to reflect.
While for reasons that have little to do with my assessment of ape-
language experiments I believe neither thesis, the continued good health
of both is evident in much of the debate over these experiments.

Terrace (Savage-Rumbaugh 1986, p. ix) contrasts the view of Descartes, that animals are 'mechanical beasts', with Darwin's opinion
that animals engage in 'forms of thinking . . . homologous to human
thought'. He then remarks that 'Until recently, there has been little
concrete basis for choosing between the contradictory positions
of Darwin and Descartes.' He continues by locating the studies of
ape-language acquisition in an incipient body of work in cognitive
psychology that is beginning to cast the balance of evidence in favour
of the Darwinian approach. While Terrace is explicitly including
the ape-language studies within a wider range of work in cognitive

psychology, language-training might seem to have particular relevance. If the question is whether animals have any thoughts at all, then perhaps the best way of finding out is to provide them with an opportunity of expressing them.

Although most researchers on the cognitive capacities of animals are, like Terrace, explicitly opposed to the Cartesian conception of animals as 'mechanical beasts', it is striking that they typically concede most of the remainder of the Cartesian perspective. Even those most ardent in defence of the cognitive achievements and capacities of non-humans frequently accept the Cartesian assumption that animal thought and even consciousness, since conceptually independent of behaviour, are in principle impossible to demonstrate (see e.g. Griffin 1984; Dawkins 1990). In ape-language research the same tendency is readily discoverable. Savage-Rumbaugh motivates her research into the variety of uses to which apes may put symbols by questioning the assumption of previous researchers that 'when an ape "correctly" uses a symbol, it had some referent clearly in mind', and thus was using the symbol to name an object (Savage-Rumbaugh 1986: 10). And while she is surely correct to argue that naming may be a complex activity, she appears not to question this Cartesian characterization of what would constitute referential use of a word. And this, surely, leaves it wholly obscure how any variety of referential uses would address the supposed issue, what the ape has 'clearly in mind'.

The untenability of these Cartesian presuppositions cannot be adequately addressed here.[10] What they require is philosophical exorcism rather than empirical research, as should perhaps be clear from the frequency with which scientists in their grip admit that, strictly speaking, no evidence could establish any of the desired conclusions about the minds of animals. For the present, I shall focus on the relation of ape-language research to some related contemporary philosophical views. I hope that this will at the same time throw some light on the inadequacy of the Cartesian perspective.

More recent philosophical views about the connection between thought and language may also seem to give particular significance to ape-language research. By contrast to Descartes's view that language

[10] The *loci classici* are Wittgenstein (1953) and Ryle (1949). The latter shows quite decisively how, even for the case of humans, a vast range of behaviour beyond the linguistic can sufficiently ground the attribution of intelligence and thought. I have tried to apply some of these insights specifically to questions about non-human mental states in Ch. 10 in this volume.

is a reliable symptom of something quite distinct from it—thought —Donald Davidson (1975, 1982) has claimed that language is a necessary condition of thought. Although, again, the arguments for and against this position cannot be thoroughly examined here,[11] ape-language research does provide a curious perspective on this claim. For if Davidson's position were correct, then what would be at issue in the ape-language controversy would be not merely whether apes could be taught to talk, but whether they could be taught to think. The idea that the apes trained by the Gardners and others differed from their wild or untrained conspecifics in that they alone possessed the rudiments of thought is so much less plausible than the idea that they have simply been taught a rudimentary form of human language as to cast doubt on the suggestion that these two contingencies might be essentially equivalent. One reason for this is that the communicative abilities of language-trained apes seem so clearly continuous with those of animals outside the educational elite. This connects with a point emphasized earlier, that even human linguistic communication is not totally different in kind from the range of non-verbal communication that provides a crucial part of its context.

At first sight a more modest proposal is that of Peter Carruthers (1989), that the thought of animals, since they lack language, must not be conscious. I say 'at first sight' more modest, because this suggestion is in fact part of a revival of the preposterous Cartesian thesis that non-human animals are not conscious at all. The considerations adduced in the previous paragraph against using language to defend such an. apartheid[12] view of human versus animal communication seem equally relevant here. Consideration of Carruthers's rationale for this neo-Cartesian (or perhaps better, neuro-Cartesian) doctrine must await another place. His discussion does, however, raise one further important issue. Questions about animal consciousness, far from constituting merely an abstruse philosophical debate, are directly relevant to an ethical issue that has recently, and appropriately, come to the forefront of philosophical debate. Specifically, a number of thinkers have recently questioned the ethical legitimacy of the prima facie horrendous treatment of animals in factory farming and scientific

---

[11] Davidson's conclusions about animals are criticized by Jeffrey (1985) and Ch. 10 in this volume.

[12] I borrow this term from an insightful paper on the meanings of animal signals by Roy Harris (1984).

research (not to mention 'recreations' such as hunting). Carruthers, on the basis of the thesis just mentioned, claims that these practices are not only permissible but, in so far as they conduce to human welfare, obligatory (Carruthers 1989: 268). While, as I hope I have made clear, I do not think the moral repugnance of this conclusion depends in any way on the success (or failure) of animal-language projects, they do, perhaps, provide useful rhetorical ammunition. Unless one adopts the absurd view that it is permissible (or obligatory) to torture any animal except an educated ape, a defender of Carruthers's position will have to defend the apartheid view of human commun-ication, with ape language falling on the animal side of this divide. I hope I have at least indicated some obstacles to this strategy.

Finally, then, there is the question whether this research has any-thing to tell us about ourselves. For the most part, this question is simply the obverse of the preceding one. That is to say, much of the official motivation for the research concerns the question whether humans are, or are not, radically discontinuous from the remainder of the animal kingdom. And the major remaining bastion for defenders of this discontinuity is the view of language as something categorially distinct from any lesser system of communication. Again, I think that this position is to be refuted not by seeing whether apes can learn our language, but rather by, first, investiga-tion of the complex and interesting lives of animals in their natural environments and, second, philosophical dismantling of the naive views of language it presupposes.

The most influential view of language that is commonly appealed to in defence of the radical-discontinuity thesis is that of Noam Chomsky (1968), which sees language as a uniquely human cognit-ive organ. But even should this (somehow) turn out to be true, it seems irrelevant to the issue under discussion. For presumably the issue concerns the kinds of things humans can do, not the organs that they may use to do them. I take it that the only serious can-didates for capacities with which a language organ might uniquely endow humans are communication and thought. But there are many kinds of non-linguistic behaviour that facilitate communica-tion, and many non-linguistic manifestations of thought (see again Ryle 1949). To argue that the existence of a special organ of language shows that only humans can think or communicate would be com-parable to arguing that only fish, being uniquely equipped with a swim-bladder, can swim.

So finally, for all its undoubted charm, I do not think the research on the linguistic aptitude of apes will tell us much about either ourselves or apes that we could not learn at least as well in many other ways. It does, however, provide the occasion for a wealth of interesting observations on the reactions and assumptions of those who engage in, or attack, this area of study. And perhaps even charm is not a totally negligible scientific virtue.

References

ATRAN, S. (1998), 'Folk Biology and the Anthropology of Science', *Behavioral and Brain Sciences*, 21: 547–611.

BARKOW, J. H., L. COSMIDES, and J. TOOBY (eds.) (1992), *The Adapted Mind* (Oxford: Oxford University Press).

BEKOFF, M., and D. JAMIESON (eds.) (1990), *Explanation and Interpretation in the Study of Animal Behavior: Comparative Perspectives* (Boulder, Colo.: Westview Press, 1990).

BENTHAM, G., and J. D. HOOKER (1926), *Handbook of the British Flora*, 7th edn., rev. A. B. Rendle (Ashford: Reeve).

BERLIN, B. D., E. BREEDLOVE, and P. H. RAVEN (1974), *Principles of Tzeltal Plant Classification* (New York: Academic Press).

BORROR, D., and R. WHITE (1970), *A Field Guide to the Insects of Northern America North of Mexico* (Boston: Houghton Mifflin).

BOURDIEU, P. (1984), *Distinction: A Social Critique of the Judgement of Taste* (Cambridge, Mass.: Harvard University Press).

BOYD, Richard (1999), 'Homeostasis, Species, and Higher Taxa', in Wilson (1999).

BOYD, Robert and P. J. RICHERSON (1985), *Culture and the Evolutionary Process* (Chicago: University of Chicago Press).

BUSS, D. (1994), *The Evolution of Desire* (New York: Basic Books).

CAPLAN, A. L. (ed.) (1978), *The Sociobiology Debate* (New York: Harper & Row).

CARRUTHERS, P. (1989), 'Brute Experience', *Journal of Philosophy*, 86: 258–69.

CARTWRIGHT, N. (1983), *How the Laws of Physics Lie* (Oxford: Oxford University Press).

CAVALLI-SFORZA, L. L., and M. W. FELDMAN (1981), *Cultural Transmission and Evolution: A Quantitative Approach* (Princeton: Princeton University Press).

CHARNOV, E. L. (1976), 'Optimal Foraging: The Marginal Value Theorem', *Theoretical Population Biology*, 9: 129–36.

CHENEY, D. L. (1984), 'Category Formation in Vervet Monkeys', in Harré and Reynolds (1984).

CHOMSKY, N. (1968), *Language and Mind* (New York: Harcourt, Brace & World).

CLARK, S. R. L. (1990), 'The Reality of Shared Emotions', in Bekoff and Jamieson (1990).

COSMIDES, L., and J. TOOBY (1987), 'From Evolution to Behaviour: Evolutionary Psychology as the Missing Link', in Dupré (1987).

CRISP, R. (1990), 'Evolution and Psychological Unity', in Bekoff and Jamieson (1990).

DANIELS, T. J., and M. BEKOFF (1990), 'Domestication, Exploitation, and Rights', in Bekoff and Jamieson (1990).

DARWIN, C. (1981), *The Descent of Man, and Selection in Relation to Sex* (Princeton: Princeton University Press).

DAVIDSON, D. (1975), 'Thought and Talk', in S. Guttenplan (ed.), *Mind and Language* (Oxford: Oxford University Press).

—— (1982), 'Rational Animals', *Dialectica*, 36: 318–27; repr. in Lepore and Mclaughlin (1985).

DAVIS, P. H. (1978), 'The Moving Staircase: A Discussion on Taxonomic Rank and Affinity', *Notes from the Royal Botanic Garden*, 36: 325–40.

DAWKINS, M. S. (1990), 'From an Animal's Point of View: Consumer Demand Theory and Animal Welfare', *Behavioral and Brain Sciences*, 13: 1–9.

DAWKINS, R. (1976), *The Selfish Gene* (Oxford: Oxford University Press).

DE LUCE, J., and H. T. WILDER (eds.) (1983), *Language in Primates* (New York: Springer).

DENNETT, D. (1971), 'Intentional Systems', *Journal of Philosophy*, 68: 87–106.

—— (1995), *Darwin's Dangerous Idea* (New York: Simon & Schuster).

DE QUEIROZ, K. (1999), 'The General Lineage Concept of Species and the Defining Property of the Species Category', in Wilson (1999).

—— and J. GAUTHIER (1990), 'Phylogeny as a Central Principle in Taxonomy: Phylogenetic Definitions of Taxon Names', *Systematic Zoology*, 39: 307–22.

—— —— (1994), 'Toward a Phylogenetic System of Biological Nomenclature', *Trends in Ecology and Evolution*, 9: 27–31.

DESCARTES, R. (1967), *Meditations on First Philosophy* (1642), in *The Philosophical Works of Descartes*, ed. and trans. E. S. Haldane and G. R. T. Ross (Cambridge: Cambridge University Press).

DOBZHANSKY, T. (1973), 'Nothing in Biology Makes Sense except in the Light of Evolution', *American Biology Teacher*, 35: 125–9.

DUPRÉ, J. (1983), 'The Disunity of Science', *Mind*, 92: 321–46.

—— (ed.) (1987), *The Latest on the Best: Essays on Evolution and Optimality* (Cambridge, Mass.: MIT Press/Bradford Books).

—— (1988), 'Materialism, Physicalism, and Scientism', *Philosophical Topics*, 16: 31–56.

—— (1993), *The Disorder of Things: Metaphysical Foundations of the Disunity of Science* (Cambridge, Mass.: Harvard University Press).

—— (1998), 'Normal People', *Social Research*, 65: 221–48.

—— (2001), *Human Nature and the Limits of Science* (Oxford: Oxford University Press).

DURHAM, W. (1978), 'Toward a Coevolutionary Theory of Human Biology and Culture', in A. Caplan (ed.), *The Sociobiology Debate* (New York: Harper & Row).

—— (1981), 'Overview: Optimal Foraging Analysis in Human Ecology', in Winterhalder and Smith (1981).

EHRLICH, P. R., and P. H. RAVEN (1969), 'Differentiation of Populations', *Science*, 165: 1228–32; repr. in Ereshefsky (1992*b*).

ELLIS, B. (1992), 'The Evolution of Sexual Attraction: Evaluative Mechanisms in Women', in Barkow *et al.* (1992).

EMLEN, J. M. (1987), 'Evolutionary Ecology and the Optimality Assumption', in Dupré (1987).

ERESHEFSKY, M. (1991), 'Species, Higher Taxa, and the Units of Evolution', *Philosophy of Science*, 58: 84–101.

—— (1992*a*), 'Eliminative Pluralism', *Philosophy of Science*, 59: 671–90.

—— (1992*b*), *The Units of Evolution: Essays on the Nature of Species* (Cambridge, Mass.: MIT Press).

—— (1999), 'Species and the Linnean Hierarchy', in Wilson (1999).

FAUSTO-STERLING, A. (1985), *Myths of Gender* (New York: Basic Books).

—— (2000), *Sexing the Body: Gender Politics and the Construction of Sexuality* (New York: Basic Books).

FLOODGATE, G. D. (1962), 'Some Remarks on the Theoretical Aspects of Bacterial Taxonomy', *Bacteriology Review*, 26: 277–91.

FOUTS, R. (1973), 'Acquisition and Testing in Four Young Chimpanzees', *Science*, 180: 978–80.

—— (1983), 'Chimpanzee Language and Elephant Tails: A Theoretical Synthesis', in de Luce and Wilder (1983).

GADGIL, M., and K. C. MALHOTRA (1983), 'Adaptive Significance of the Indian Caste System: An Ecological Perspective', *Annals of Human Biology*, 10: 465–78.

GARDNER, B., and A. GARDNER (1971), 'Two-Way Communication with an Infant Chimpanzee', in A. M. Schrier and F. Stollnitz (eds.), *Behavior of Non-Human Primates* (New York: Academic Press, 1971), iv.

GAUTHIER, D. (1986), *Morals by Agreement* (Oxford: Oxford University Press).

GHISELIN, M. (1974), 'A Radical Solution to the Species Problem', *Systematic Zoology*, 23: 536–44.

—— (1987), 'Species Concepts, Individuality, and Objectivity', *Biology and Philosophy*, 2: 127–43.

—— (1997), *Metaphysics and the Origin of Species* (Albany, NY: State University of New York Press).

GILMOUR, J. S. L. (1951), 'The Development of Taxonomic Theory since 1851', *Nature*, 168: 400–2.

GODFREY-SMITH, P. (1994), 'A Modern History Theory of Functions', *Nous*, 28: 344–62.

—— (2000), 'On the Theoretical Role of "Genetic Coding"', *Philosophy of Science*, 67: 26–44.

GOLDBERG, S. (1973), *The Inevitability of Patriarchy* (New York: William Morrow).

GORDON, D. M. (1997), 'The Genetic Structure of *Escherichia coli* Populations in Feral Mice', *Microbiology*, 143: 2039–46.

GOULD, S. J., and R. C. LEWONTIN (1979), 'The Spandrels of San Marco and the Panglossian Paradigm: A Critique of the Adaptationist Programme', *Proceedings of the Royal Society of London*, 205: 581–98.

GRIFFIN, D. R. (1984), *Animal Thinking* (Cambridge, Mass.: Harvard University Press).

GRIFFITHS, P. E., and R. D. GRAY (1994), 'Developmental Systems and Evolutionary Explanations', *Journal of Philosophy*, 91: 277–304.

GYLLENBERG, H. G., M. GYLLENBERG, T. KOSKI, T. LUND, J. SCHINDLER, and M. VERLAAN (1997), 'Classification of *Enterobacteriacea* by Minimization of Stochastic Complexity', *Microbiology*, 143: 721–32.

HACKING, I. (1983), *Representing and Intervening* (Cambridge: Cambridge University Press).

—— (1999), *The Social Construction of What?* (Cambridge, Mass.: Harvard University Press).

HARRÉ, R., and V. REYNOLDS (eds.) (1984), *The Meaning of Primate Symbols* (Oxford: Oxford University Press).

HARRIS, R. (1984), 'Must Monkeys Mean?', in Harré and Reynolds (1984).

HEDIGER, H. (1980), 'Do you Speak Yerkish? The Newest Colloquial Language with Chimpanzees', in Sebeok and Umiker-Sebeok (1980).

HILL, J. H. (1980), 'Apes and Language', in Sebeok and Umiker-Sebeok (1980).

HOLMSTROM, N. (1982), 'Do Women have a Distinct Nature?', *Philosophical Forum*, 14: 25–42.

HRDY, S. B. (1981), *The Woman who never Evolved* (Cambridge, Mass.: Harvard University Press).

HULL, D. L. (1965), 'The Effect of Essentialism on Taxonomy: Two Thousand Years of Stasis', *British Journal for the Philosophy of Science*, 15: 314–26, 16: 1–18; repr. in Ereshefsky (1992*b*).

—— (1976), 'Are Species Really Individuals?', *Systematic Zoology*, 25: 174–91.

—— (1989), *The Metaphysics of Evolution* (Albany, NY: State University of New York Press).

JAGGAR, A. (1983), *Feminist Politics and Human Nature* (Totowa, NJ: Rowman & Allanheld).

JEFFREY, R. (1985), 'Animal Interpretation', in Lepore and McLaughlin (1985).

KAUFFMAN, S. (1993), *The Origins of Order* (New York: Oxford University Press).

—— (1995), *At Home in the Universe* (New York: Oxford University Press).

KEENE, A. (1981), 'Optimal Foraging in a Nonmarginal Environment', in Winterhalder and Smith (1981).

KITCHER, P. (1982), 'Genes', *British Journal for Philosophy of Science*, 33: 337–59.

—— (1984), 'Species', *Philosophy of Science*, 51: 308–33; repr. in Ereshefsky (1992*b*).

—— (1985), *Vaulting Ambition: Sociobiology and the Quest for Human Nature* (Cambridge, Mass.: MIT Press).

—— (1987), 'Why Not the Best?', in Dupré (1987).

KITTS, D. B., and D. J. KITTS (1979), 'Biological Species as Natural Kinds', *Philosophy of Science*, 46: 613–22.

KRIPKE, S. (1972*a*), 'Identity and Necessity', in M. Munitz (ed.), *Identity and Individuation* (New York: New York University Press).

—— (1972*b*), 'Naming and Necessity', in D. Davidson and G. Harman (eds.), *Semantics of Natural Language* (Dordrecht: Reidel).

KUHN, T. S. (1970), *The Structure of Scientific Revolutions*, 2nd edn. (Chicago: Chicago University Press).

LAVIN, M. (unpublished), 'On the Moral Permissibility of Sexual Reassignment Surgery'.

LEHMAN, H. (1967), 'Are Biological Species Real?', *Philosophy of Science*, 34: 157–67.

LEPORE, E., and B. T. McLAUGHLIN (eds.) (1985), *Actions and Events: Perspectives on the Philosophy of Donald Davidson* (Oxford: Blackwell).

LEWONTIN, R. C. (1987), 'The Shape of Optimality', in Dupré (1987).

—— and L. C. DUNN (1960), 'The Evolutionary Dynamics of a Polymorphism in the House Mouse', *Genetics*, 45: 705–22.

——, S. ROSE, and L. KAMIN (1984), *Not in our Genes: Biology, Ideology, and Human Nature* (New York: Pantheon).

LOCKE, J. (1975), *An Essay Concerning Human Understanding* (1689), ed. P. H. Nidditch (Oxford: Oxford University Press).

LONGINO, H., and R. DOELL (1983), 'Body, Bias, and Behavior: A Comparative Analysis of Reasoning in Two Areas of Biological Science', *Signs*, 9: 206–27.

LORENZ, K. (1977), *Behind the Mirror: A Search for a Natural History of Human Knowledge* (London: Methuen).

MAGNUS, D. (1998), 'Evolution without Change in Gene Frequencies', *Biology and Philosophy*, 13: 255–61.

MAY, R. M. (1973), *Stability and Complexity in Model Ecosystems* (Princeton: Princeton University Press).

MAYNARD SMITH, J. (1978), *The Evolution of Sex* (Cambridge: Cambridge University Press).

MAYR, E. (1961), 'Causes and Effect in Biology', *Science*, 134: 1501–6.

—— (1963), *Animal Species and Evolution* (Cambridge, Mass.: Harvard University Press).

—— (1975), *Populations, Species, and Evolution* (Cambridge, Mass.: Harvard University Press).

MAYR, E. (1987), 'The Ontological Status of Species: Scientific Progress and Philosophical Terminology', *Biology and Philosophy*, 2: 145–66.

—— (1996), 'What is a Species, and What is Not', *Philosophy of Science*, 63: 262–77.

—— (1997), *This is Biology: The Science of the Living World* (Cambridge, Mass.: Harvard University Press).

MILES, H. L. (1983), 'Apes and Language: The Search for Communicative Competence', in de Luce and Wilder (1983).

MILL, J. S. (1862), *A System of Logic*, 5th edn. (London: Parker, Son, & Bourn).

MISHLER, B. D. (1999), 'Getting Rid of Species?', in Wilson (1999).

—— and R. N. BRANDON (1987), 'Individuality, Pluralism, and the Biological Species Concept', *Biology and Philosophy*, 2: 397–414.

—— and M. J. DONOGHUE (1982), 'Species Concepts: A Case for Pluralism', *Systematic Zoology*, 31: 491–503; repr. in Ereshefsky (1992*b*).

MORGAN, E. (1972), *The Descent of Woman* (London: Corgi).

MORRIS, D. (1967), *The Naked Ape* (New York: McGraw-Hill).

NAGEL, T. (1974), 'What is it Like to be a Bat?', *Philosophical Review*, 83: 435–50.

NANNEY, D. L. (1980), *Experimental Ciliatology* (New York: John Wiley).

—— (1999), 'When is a Rose? The Kinds of *Tetrahymena*', in Wilson (1999).

NICHOLSON, B. E., S. ARY, and M. GREGORY (1960), *The Oxford Book of Wildflowers* (Oxford: Oxford University Press).

NIKLAS, K. J. (1997), *The Evolutionary Biology of Plants* (Chicago: University of Chicago Press).

OPPENHEIM, P., and H. PUTNAM (1958), 'The Unity of Science as a Working Hypothesis', in H. Feigl and M. Scriven (eds.), *Minnesota Studies in the Philosophy of Science*, ii (Minneapolis: University of Minnesota Press).

ORTNER, S., and H. WHITEHEAD (1981), *Sexual Meanings* (Cambridge: Cambridge University Press).

OYAMA, S. (1985), *The Ontogeny of Information* (Cambridge: Cambridge University Press).

PACE, N. R. (1997), 'A Molecular View of Microbial Diversity and the Biosphere', *Science*, 276: 734–40.

PATTERSON, F., and E. LINDEN (1981), *The Education of Koko* (New York: Holt, Rinehart & Winston).

PREMACK, A. (1976), *Why Chimps Can Read* (New York: Harper & Row).

PREMACK, D. (1972), 'Teaching Language to an Ape', *Scientific American*, 227: 92–9.

—— (1986), *Gavagai* (Cambridge, Mass.: Bradford Books/MIT Press).

PULLIAM, H. R., and C. DUNFORD (1980), *Programmed to Learn: An Essay on the Evolution of Culture* (New York: Columbia University Press).

PUTNAM, H. (1975*a*), 'Explanation and Reference', in Putnam (1975*d*).

—— (1975*b*), 'Is Semantics Possible?', in Putnam (1975*d*).

—— (1975c), 'The Meaning of "Meaning"', in Putnam (1975d).

—— (1975d), *Mind, Language, and Reality, Philosophical Papers*, ii (Cambridge: Cambridge University Press).

QUIATT, D. (1984), 'Devious Intentions of Monkeys and Apes', in Harré and Reynolds (1984).

QUINE, W. (1969), 'Natural Kinds', in Quine, *Ontological Relativity and Other Essays* (New York: Columbia University Press).

REED, E. (1978), *Sexism and Science* (New York: Pathfinder).

RICHERSON, P. J., and R. BOYD (1987), 'Simple Models of Complex Phenomena: The Case of Cultural Evolution', in Dupré (1987).

ROSENBERG, A. (1980), *Sociobiology and the Preemption of Social Science* (Baltimore: Johns Hopkins University Press).

ROSENZWEIG, M. L. (1990), 'Do Animals Choose Habitats?', in Bekoff and Jamieson (1990).

RUMBAUGH, D. (1977), *Language Learning by a Chimpanzee: The Lana Project* (New York: Academic Press).

RUSE, M. (1969), 'Definitions of Species in Biology', *British Journal for the Philosophy of Science*, 20: 97–119.

—— (1987), 'Biological Species: Natural Kinds, Individuals, or What?', *British Journal for the Philosophy of Science*, 38: 225–42; repr. in Ereshefsky (1992b).

RYLE, G. (1949), *The Concept of Mind* (London: Hutchinson).

SALMON, N. (1981), *Reference and Essence* (Princeton: Princeton University Press).

SAVAGE-RUMBAUGH, S. (1986), *Ape Language* (New York: Columbia University Press).

SCHAUER, T. (1982), *A Field Guide to the Wild Flowers of Britain and Europe*, trans. R. Pankhurst (London: Collins).

SCHLESINGER, G. (1963), *Method in the Physical Sciences* (New York: Humanities Press).

SEBEOK, T. A. (1980), 'Looking in the Destination for what should have been Sought in the Source', in Sebeok and Umiker-Sebeok (1980).

—— and UMIKER-SEBEOK, J. (eds.) (1980), *Speaking of Apes* (New York: Plenum).

SEIDENBERG, M. S., and PETITTO, L. A. (1979), 'Signing Behavior in Apes: A Critical Review', *Cognition*, 1: 177–215.

SEYFARTH, R. M. (1984), 'What the Vocalizations of Monkeys Mean to Humans and what they Mean to the Monkeys Themselves', in Harré and Reynolds (1984).

SHEPARD, R. N. (1987), 'Evolution of a Mesh between Principles of the Mind and Regularities of the World', in Dupré (1987).

SINGH, D. (1993), 'Adaptive Significance of Waist-to-Hip Ratio and Female Physical Attractiveness', *Journal of Personality and Social Psychology*, 65: 293–307.

SKINNER, B. F. (1953), *Science and Human Behavior* (New York: Macmillan).

SMITH, E. A. (1981), 'Optimal Foraging Theory and the Analysis of Hunter-Gatherer Group Size', in Winterhalder and Smith (1981).

—— (1987), 'Optimization in Anthropology: Applications and Critiques', in Dupré (1987).

SNEATH, P. H. A., and R. R. SOKAL (1973), *Numerical Taxonomy* (San Francisco: W. H. Freeman).

SOBER, E. (1984*a*), 'Discussion: Sets, Species, and Evolution: Comments on Philip Kitcher's "Species"', *Philosophy of Science*, 51: 334–41.

—— (1984*b*), *The Nature of Selection* (Cambridge, Mass.: MIT Press/ Bradford Books).

—— (1987), 'What is Adaptationism?', in Dupré (1987).

—— (1992), 'Monophyly', in E. A. Lloyd and E. F. Keller (eds.), *Keywords in Evolutionary Biology* (Cambridge, Mass.: Harvard University Press).

—— (1994), *From a Biological Point of View* (Cambridge: Cambridge University Press).

—— and D. S. WILSON (1998), *Unto Others: The Evolution and Psychology of Unselfish Behavior* (Cambridge, Mass.: Harvard University Press).

SOKAL, R., and P. SNEATH (1963), *Principles of Numerical Taxonomy* (San Francisco: W. H. Freeman).

SPELLENBERG, R. (1969), *The Audubon Society Field Guide to North American Wildflowers: Western Region* (New York: Alfred A. Knopf).

STEBBINS, G. L. (1987), 'Species Concepts: Semantics and Actual Situations', *Biology and Philosophy*, 2: 198–203.

STERELNY, K. (1999), 'Species as Ecological Mosaics', in Wilson (1999).

TEMPLETON, A. R. (1989/1992), 'The Meaning of Species and Speciation: A Genetic Perspective', in D. Otte and J. A. Endler (eds.), *Speciation and its Consequences* (Sunderland, Mass.: Sinauer, 1989); repr. in Ereshefsky (1992*b*).

TERRACE, H. (1987), *Nim* (New York: Columbia University Press).

THORNHILL, R., and N. W. THORNHILL (1983), 'Human Rape: An Evolutionary Analysis', *Ethology and Sociobiology*, 4: 137–73.

—— —— (1992), 'The Evolutionary Psychology of Men's Coercive Sexuality', *Behavioral and Brain Sciences*, 15: 363–421.

UMIKER-SEBEOK, J., and T. A. SEBEOK (1980), 'Questioning Apes', in Sebeok and Umiker-Sebeok (1980).

VANDAMME, P., B. POT, M. GILLIS, P. DE VOS, K. KERSTERS, and J. SWINGS (1996), 'Polyphasic Taxonomy: A Consensus Approach to Bacterial Taxonomy', *Microbiological Review*, 60: 407–38.

VAN VALEN, L. (1976), 'Ecological Species, Multispecies, Oaks', *Taxon*, 25: 233–9; repr. in Ereshefsky (1992*b*).

WALTERS, S. M. (1961), 'The Shaping of Angiosperm Taxonomy', *New Phytologist*, 60: 74–84.

WILLIAMS, G. C. (1975), *Sex and Evolution* (Princeton: Princeton University Press).

WILLIS, J. C. (1949), *Birth and Spread of Plants* (Geneva: Conservatoire et Jardin Botanique de la Ville).

WILSON, E. O. (1975), *Sociobiology: The New Synthesis* (Cambridge, Mass.: Harvard University Press).

—— (1978), *On Human Nature* (Cambridge, Mass.: Harvard University Press).

WILSON, R. A. (ed.) (1999), *Species: New Interdisciplinary Essays* (Cambridge, Mass.: MIT Press/Bradford Books).

WINTERHALDER, B. (1981), 'Optimal Foraging Strategies and Hunter-Gatherer Research in Anthropology', in Winterhalder and Smith (1981).

—— and E. A. SMITH (eds.) (1981), *Hunter-Gatherer Foraging Strategies* (Chicago: University of Chicago Press).

WITTGENSTEIN, L. (1953), *Philosophical Investigations*, trans. G. E. M. Anscombe (Oxford: Blackwell).

WOESE, C. R. (1987), 'Bacterial Evolution', *Microbiological Review*, 51: 221–71.

Index